D1278121
TOLEDO
ORGANIC PRAIRIE
Family of Farms
ORGANIC FARM
NO
SPRAY
ORGANIC VALLEY
ORGANIC VALLEY
¡VIVA
LA
PASTURE!
WOW!
AMAZING
&
HUGE
GARLIC !!!

CROPP Cooperative ROOTS

The First 25 Years

We dedicate this book
to the idea that communal action can
preserve the organic integrity of our earth.

Printed in Canada

International Standard Book Number
ISBN: 978-0-9887748-0-3

Library of Congress Control Number: 2007012345

The interior of this book is printed on 100% post-consumer recycled paper that is acid free and processed
chlorine free, using vegetable-based inks. The paper is certified by the Forest Stewardship Council (FSC).

* * *

*Handwritten elements throughout the book were penned by Jerome McGeorge,
CROPP's original CFO, who hand-wrote the Cooperative's earliest annual reports.*

When I was in grade school in the late 1960s and early 1970s, the school bus picked up every kid in our area of Vernon County, Wisconsin. I would say 90% of the kids were dairy farmers. When my kids went to school in the 1980s, they were the only dairy farmers that rode their school bus. Everybody else had stopped dairying by then. The worst thing is it's all the small farms that are gone now. That's one thing CROPP Cooperative is doing; we're saving those small farms.

CROPP Cooperative Roots highlights the Cooperative's accomplishments in the last 25 years, to show people that it can be done. It is possible to be fair to consumers, and fair to the farmers who are producing the products.

Can you imagine a farmer who isn't farming organically picking up this book and reading about the history and what the Co-op's done? They would say, "Man, I'd like to be part of that co-op." I could see that happening.

Can you imagine a mother who cares deeply about her family's future reading this book and what the Co-op stands for? She would say, "Wow, I'd love to provide food with this much integrity to my family." I could see that happening too.

I feel good about being part of CROPP. We've had good people on the management team who know what it takes to farm and we've had a good board of directors. It seems like we've made a lot of sound decisions over these 25 years.

I've got two sons who are part of CROPP, too. One actually works as a CROPP employee and grows produce and one's dairying. It's satisfying to have another generation remain on the farm.

There can't be a better feeling than to know we're changing so many farmers' and consumers' lives.

Sincerely,

Arnold J. Trussoni

Arnie Trussoni
CROPP Board President

FARMY

"From the barn whose vaned cupola was visible over the house roof against the pale sky, Mat Feltner was calling his cows. Old Jack (Beecham) listened with an eagerness that carried him away from himself; for all his consciousness of where he was, he might have been asleep and dreaming. Mat waited, and called again. And then from the quietening of Mat's voice, Old Jack knew that the cows had come near and that Mat could see them moving up deliberative and shadowy out of the mists and the thinning darkness. And then he heard the barn doors slide open." Wendell Berry, *The Memory of Old Jack*

Why are young people committing their lives to running the family farm when, just a generation ago, their parents would have warned them to leave the farm and get decent jobs?

Why has the organic way of farming, dating back to the dawn of civilization but deemed archaic by the 1960s, become the fastest-growing sector of the American food market in the last two decades?

A rare "harmonic convergence" of forces brought this about—each a ripple in an ever-larger pool of human experience.

For a group of frustrated farmers in southwest Wisconsin, the forces were powerfully personal. One farm family lost its grandfather—the sole handler of insecticides and pesticides for years—to cancer. Another watched 13 cows die when they ate insecticide-tainted hay. A third farmer simply longed to hug his children when he came in from chores, but was denied by the fearsome toxins that clung to him instead. Though farming was America's most deeply rooted way of life, it was increasingly dangerous, and it could no longer promise a decent living. Family farmers—admonished to "get big or get out" by their government—carried massive debt, while the value of their farm land and their production plummeted in the 1980s.

It might have been far easier to give up and get out, but these farmers refused. They dared to believe that family farms could survive the economic crisis that plagued them. They could even thrive by farming differently.

These farmers were convinced that human health and the health of the environment and the soil were all connected. Some people called them pioneers, but others labeled them cranks and starry-eyed idealists. How could a handful of stubbornly committed farmers ever swim against the current of modern, chemical-based farming to produce better food?

They were on their own, or so it seemed . . . until they discovered a torrent of kindred spirits across America, joined hands with them and succeeded.

The Roots of Today's Organic Movement

Observers say the American organic movement had its kick start in 1969 at the University of California when students calling themselves the Robin Hood Commission seized a vacant lot and turned it into a garden for "uncontaminated" vegetables. They called it the People's Park. (This was decades before obesity, cancer, diabetes and heart disease were scientifically linked to the kinds of food we eat.)

Truth is, the modern organic movement's roots went far deeper, into the earliest years of the last century.

In the 1960s, Jerome Rodale—an electrical supply dealer turned organic farmer—also saw sales of his *Organic Gardening and Farming* magazine take off when America's counter-culture magazine, the *Whole Earth Catalog,* gave his publication a rave review. Rodale had been publishing since the 1940s with little fanfare and few readers (Wendell Berry was one of its authors). Rodale's back-to-the-land organic sensibility appealed to many of America's youth who were anti-war, anti-big business and anti-social convention.

> "One of these days, the public is going to wake up and will pay for eggs, meat and vegetables according to how they were produced." Jerome Rodale, founder of Rodale, Inc. May, 1942

Rodale had also published *Pay Dirt* in 1945, a book that underscored the dangers of arsenic sprays and the newly-developed DDT. He offered a bold alternative to chemicals by espousing the organic principles developed by his friend and mentor Sir Albert Howard, a British agronomist who is considered the father of modern organic agriculture.

Born and raised on a farm and having experimented with composting on India farm land since 1905, Sir Howard advocated using plant and animal wastes to make compost to replenish the earth, making the soil rich and healthy. He called this the "Law of Return," and he described it in his 1940 book, *An Agricultural Testament.*

Fellow Brits, many of whom were agricultural innovators, activists and authors in Europe, endorsed Howard's thinking. Across the pond, Rodale was Howard's apostle, but the opposing practitioners of modern agriculture were more powerful in the United States.

NPK: Scientific and Certain

American universities promoted a form of agriculture they considered "more scientific" that was based heavily on the work of German chemist Baron Justus Liebig who, 100 years earlier, had identified three chemical elements that plants need to grow: nitrogen, phosphorus and potassium. Take those elements (called NPK for their initials on the periodic table), combine them into fertilizer, mix with the soil, plant crops and—voila!—boost yields. Simple, scientific, certain.

Synthesized fertilizer, gas-powered tractors and chemical control of pests became the norm in the United States beginning around the 1930s. So did debt for farmers who were later admonished by President Richard Nixon's Secretary of Agriculture, Earl Butz, to "get big . . . or get out." Successful farming was all about more of everything: more chemicals, more machinery, higher yields, abundant (and cheaper) food.

Could it continue? America's policy-makers thought so.

Contrary Voices

Now flash back to the People's Park in 1969, where the students sold their "uncontaminated" food and sang the praises of Rodale and organic purity. Many had read *Silent Spring,* a book authored in 1962 by Rachel Carson, a biologist from the U.S. Bureau of

Fisheries who explained in plain, passionate language the dangers of synthetic pesticides and the impact that humans have on our environment. Her work ignited the start of the environmental movement in America during the 1960s and 1970s. (Remember our first Earth Day? It was April 22, 1970, created by Wisconsin Senator Gaylord Nelson.)

In 1971, Frances Moore Lappé wrote a bestseller that got people talking. In *Diet for a Small Planet,* she argued that world hunger was caused not by a lack of food, but by ineffective food policy, including wasteful, grain-fed meat production. She offered simple rules for a healthy, wholesome, balanced diet that didn't require meat. By that time, Adelle Davis had also become America's best known nutritionist, advocating a diet of unprocessed foods while blasting the food industry in her book, *Let's Eat Right To Keep Fit.* In that same lively decade, muscle-man Jack LaLanne—an admitted former junk food addict—became America's "father of fitness" on network TV, preaching the health benefits of regular exercise, a good diet and blaming overly-processed foods for many health problems.

Talk about food—how it was grown, where it came from, how it was processed and what it cost—escalated in the 1970s when scientific evidence revealed the danger of pesticides and a world oil crisis threatened to put the brakes on modern agriculture. Maybe there was an alternative way to grow food. Maybe the time was ripe for a return to harmony between farmer, consumer and earth.

Back to the Land

Mr. Rodale's readers of *Organic Gardening and Farming* swelled to 700,000 in 1971. Some of those readers were avid, backyard organic gardeners and supporters of the new organic food co-ops and fledgling community supported agriculture (CSA) buying groups. Some of them took their college degrees and left the convenience of urban living behind. They were high on rural romance and the excitement of doing agriculture one better. They went back to the land to create alternative lives, gathering in groups to live together and farm. Most had little or no experience.

> **"The rural communes served as organic agriculture's ramshackle research stations, places where neophyte farmers could experiment . . . "** Michael Pollan, *The Omnivore's Dilemma,* 2006

There was even talk and some action in a few states focused on certifying organic farmers so that consumers could be sure they were buying the real deal.

Of course, the "back-to-the-landers" weren't the only farmers using organic farming methods. Some veteran farmers never embraced the notions of big, chemical agriculture. Many looked to their faith for guidance. Others were simply too cheap or too practical to pay for expensive and not always reliable synthetics and chemicals to boost their crop production, kill insects, fatten their cattle, treat livestock disease and increase their dairy production. They followed the time-honored practices learned from their parents and grandparents. Looking back, one wizened organic farmer admitted, "I'm so far behind, now I'm ahead."

> **"We abuse land because we regard it as a commodity belonging to us. When we see land as a community to which we belong, we may begin to use it with love and respect."** Aldo Leopold, Wisconsin ecologist and author

The Boom, Then the Bust

American agriculture enjoyed an all-out boom in the 1970s, whether people farmed the modern, industrial way or used old-fashioned organic methods. The Soviet Union made record purchases of American grain and—due to relaxation of trade barriers combined with heavy U.S. subsidies—most agricultural exports took off. Farm commodity prices rose and so did farm incomes. Land values spiked, too, and interest rates were seductively low. Farmers found it easy to borrow.

But all that optimism and faith in the future was snuffed out in the early 1980s. Interest rates climbed to double digits, farmers overproduced, commodity prices dropped and the agriculture boom went bust. By 1984, farm debt reached a record $215 billion—15 times what it was in 1950. Land values plummeted and many farmers could no longer carry their debt because the interest payments exceeded their incomes. Rural banks teetered and small town businesses that depended on farming seized up.

> "I was raised to believe that the farmer is the backbone of the country and that farming is the bottom rung on the economic ladder. When that agricultural rung is broken or sprung, everything above it goes haywire."
> Willie Nelson, 2002

Critics said President Ronald Reagan's farm policy was ad hoc and ineffective, creating "anarchy in the marketplace." They argued that subsidies intended to keep people on their farms were largely going to the fewest and biggest farms, not the little guys after all.

The Federal Reserve Board estimated that more than one-third of America's farms were in big trouble in the early to mid-1980s. Many lost their farms—"home places" dating back four and five generations. Graphic news stories described suicides, alcoholism and child abuse among some of America's stricken farm families. As newsman and commentator Hugh Sidey described it, "The farm economic crisis has become . . . a cultural crisis unique in our history."

Then, the Revolution

The idea for a first-ever Farm Aid concert to help American farmers in danger of losing their farms materialized on September 22, 1985. The founding fathers of that concert—Willie Nelson, John Mellencamp and Neil Young—raised more than $9 million when a crowd of 80,000 people packed the University of Illinois Memorial Stadium. Nelson and Mellencamp later brought farmers to testify before Congress, and their efforts helped pass the Agricultural Credit Act of 1987. It was expressly designed to help save family farms from foreclosure. Farm Aid is still active today.

Economic calamity in the rural heartland and a growing awareness of the problems with agricultural chemicals started a revolution. The winds of change were beginning to blow and the first barely discernible breeze stirred the imaginations of a small group of committed farmers in an overlooked corner of Wisconsin.

These farmers dreamed of a cooperative venture that would give family farmers a way to stay on the land they loved *and* give consumers better food, produced without dangerous chemicals. They knew their idea was a good one, but they never dreamed it would grow to become the largest organic farmer-owned cooperative in the United States.

1. The Best and

1975 - 1985

the Worst of Times

* Farmers are blessed with common sense * None of us are as

smart as all of us * Dynamic tension is a natural state of effectiveness *

"There is no neat and easy way to tell the story of a farm," Wisconsin author Ben Logan wrote in *The Land Remembers,* first published in 1975. "A farm is a process, where everything is related, everything happening at once. It is a circle of life."

Logan captured a year in the life of his famly's 260-acre farm standing on one of many hilltops bordered by woods and high ridges overlooking the wild and winding Kickapoo River Valley in southwest Wisconsin. It is a region that puzzles geologists to this day because the Ice Age glaciers drifted *around* it, leaving the undulating landscape intact instead of raked and flattened.

They call this region that covers the Upper Mississippi River Basin—including parts of Wisconsin, Minnesota, Iowa and Illinois—"the Driftless Region," and nowhere is this identity more apparent than Vernon and Crawford counties of southwest Wisconsin. People sport Driftless t-shirts, coffee mugs, tattoos and bumper stickers. There is considerable pride in successful farming here because, though the soil is deep and black and pristine springs are abundant, there is no place for a large spread. The land is sculpted and forested with many valleys cut into limestone. Those hills are untamable and the waters virtually unnavigable.

The farmers who first settled here generations ago, followed by the Amish and a new breed of organic farmers in the 1970s, had to find alternatives to, "get big or get out." They had to be resourceful, learn from each other and work together. They had to be creative and entrepreneurial if they were to make their lives atop these scenic ridges or tucked into protected coulees.

"It's a magical place where people live for and with the land," one young descendent of a Vernon County organic farmer observed. He echoes what Ben Logan said of his own experience: "Once you have lived on the land, been a partner with its moods, secrets and seasons, you cannot leave. The living land remembers, touching you in unguarded moments, saying, 'I am here. You are part of me.'"

Even when America's farmers faced economic calamity in the 1980s, the people you are about to meet chose not to leave the land. Their commitment to stay and find a better way to farm begins this story of perseverance, sacrifice and risk.

> "I cannot leave the land. How can I when a thousand sounds, sights and smells tell me I am part of it?" Ben Logan, *The Land Remembers*

Farmers are Blessed with Common Sense

Wayne Peters is a fourth generation Wisconsin native in his late 70s who, along with his three sons, Rory, Roger and Daniel, farm about 650 acres with dairy cows, pigs and chickens. Their operation is all organic and they've been farming as a foursome for more than 32 years on a high ridge north of Chaseburg, Wisconsin.

Peters and his wife, Irene, have lived and worked on this land since 1959. Peters' father, grandfather and great grandfather all farmed their home place near La Crosse, Wisconsin. He is a no-nonsense, practical guy who grew up believing that the only way to farm was to use chemicals and commercial fertilizers. As a young man supplementing his farm income, he sold anhydrous ammonia, a widely-used source of nitrogen fertilizer that, due to its chemical properties, is potentially one of the most dangerous chemicals used in agriculture.

Universities were singing the praises of chemicals and commercial fertilizers in the 1950s, Peters says, but he admits he was "inherently cheap and didn't want to buy fertilizer and spend money he didn't have to" on his own farm. He gravitated to more organic farming when he began selling farm implements and silos for two

← View from the Trussoni farm in Vernon County, Wisconsin

← Wayne Peters' farm, 1980s (left); the Peters family and neighbors including Ray Hass (seated, center), 1980s

Allan Haas with his daughters Katie (left) and Tammy, 1993 →

companies whose people taught him "how to get the most out of what I already had. I began learning to take care of my land in the right way."

The farm Peters bought in Vernon County got its start as a model of a new farming practice in the 1930s called contour stripping. It preserved the soil from erosion and captured rainwater before it escaped down the farm's hilly terrain. Peters plowed manure into the soil year after year, instead of relying on expensive chemical fertilizers. As Peters acquired more farm land, he worked to restore the starved soil, building it up with nutrient-rich humus. By the 1980s, he had learned enough about organic farming practices to know that it could save him money and preserve his soil. Never mind that people called him a "generic" farmer because they couldn't quite get the term right. Peters even hosted a sustainable farming meeting in his home to share ideas and maybe create a few organic converts. He counted on the fact that farmers are richly blessed with common sense. When they saw the benefits of organic farming, they might join him.

Ray's Epiphany

Peters' friend Ray Hass was one of those common sense dairy farmers whose family ownership of land in Harmony Township, Wisconsin, dated back to 1885. He was steeped in land stewardship and had veered away from conventional farming by the early 1970s. Hass and his son, Allan, had built a manure storage pit and they preferred natural fertilizers and manure over synthetic substitutes. But Hass had the ultimate epiphany when he sprayed a popular herbicide, Atrazine, on his corn crop. Some of that spray drifted into Hass' nearby alfalfa crop and killed it. "That was it," says Allan, who became a 50-50 farmer-owner with his dad in 1985. "Dad was done with chemicals."

The younger Hass was the target of ribbing in high school during the 1970s when he focused his class project on organic agriculture. "'That's stupid, you can't do organic,' the kids said, but I had an open-minded teacher who encouraged me." Hass would eventually marry his wife, Lisa Dickman, raise three daughters and commit his career to organic family farming.

No question about it, organic farmers in Vernon County, Wisconsin, were a distinct minority in the 1970s, but the pendulum started to swing in the 1980s.

Smart or a Screw Loose?

A former school teacher, Dean Swenson, was one of the early-adopters of organic in 1980, and his wife, Jan, recalls the neighbors' reactions vividly. "We had a very close-knit community and all the neighbors thought Dean was crazy," says Jan. "No one else we knew farmed organically. Everyone respected Dean, but they thought he had a screw loose. They were waiting to see him fail."

Swenson had assumed responsibility for the family

home place at age 32 and, though he thought he was too old for farming, his wife encouraged him. The Swensons had been conventional dairy farmers through the 1970s, but a trip to Germany gave them a new perspective. "We visited a 'biodynamic' farm and I came back excited about organic," Dean says. Next, they met a Wisconsin couple engaged in organic farming and learned more. When the Swensons attended a program on world hunger in Minneapolis, they got to know a Minnesotan who had farmed organically for 26 years. They were inspired.

Conventional farming wasn't doing Swenson any favors. Sometimes spraying his corn with chemicals killed his crop. Though farm prices had increased through the 1970s, so had the costs of synthetic fertilizers, herbicides and pesticides. Swenson's profit margin was miniscule.

"I called the University of Wisconsin to ask if they had any information on how to start organic farming and the person I talked to laughed," Swenson recalls. "He said he could dig up something from back in the 1930s if I wanted it. The University was totally in favor of chemical farming."

"Dean went cold turkey in 1980," says Jan. "No chemicals. It was radical. Our neighbors drove by to see how our corn looked. Little by little, they began to realize that this organic approach was working. We started having 'field days' and before we knew it, our whole neighborhood was farming organically."

Help Thy Neighbor

When Jim Wedeberg's grandfather, John, emigrated from Norway to Crawford County, Wisconsin, in 1890, he began farming the same land that Jim and Julie Wedeberg and their sons, John and Jake, farm today. They are in their fifth generation of farming.

Jim Wedeberg was 29 when he bought the home place in 1980 and his mortgage rate was 8 percent. Within three years, that figure spiked to 17.25 percent during the economic recession of the early 1980s and the start of the American farm crisis. If he was lucky, Wedeberg could pay the interest, but little more. "There was so much pain and loss in the countryside," he says, "so many foreclosures. Agriculture hit the worst of times. We all felt it."

Along with dairy, many farms in Vernon and Crawford counties grew top grade tobacco for chewing and hand-wrapped premium cigars. It was a tradition dating back to the 1880s because of the region's rich soil. But that labor-intensive crop was in serious decline in the early 1980s because manufacturers had turned to countries with cheaper labor that offered lesser quality tobacco at lower prices.

Wedeberg joined a group of farmers who sponsored local meetings to help their struggling neighbors. "It was like a support group," he says. "We invited people to come and we'd look for ways we could work together to help each other." The issues were deeply personal,

Dean and Jan Swenson, 1983; Jim Wedeberg on his dairy farm, 1995; Arnie Trussoni with his wife, Tama, 1979 →

emotional and practical. It was a loosely-organized group motivated by compassion. "All we had was time," Wedeberg says, "and we tried to put it to good use. Looking back, we felt we were really doing something, though we were all struggling ourselves."

Fairness for Farmers

Adolph Trussoni, a second generation dairy and tobacco farmer, likes a good cigar as much as anyone, but he remembers how the quality took a nosedive when manufacturers turned their backs on Wisconsin's tobacco producers. Trussoni's farm in Vernon County dates back to 1875, and his father bought the spread in 1945. Largely because of cost, the family didn't use commercial fertilizer and only a light application of Atrazine for their corn, a herbicide that later proved to be dangerous. Trussoni remembers watching an Atrazine salesman pitch his product to local farmers claiming that it was so safe he could drink it. He chugged the beaker of liquid and Trussoni was incredulous. Whether it was the real thing, no one knows.

Between 1980 and 1988, consumer food prices rose 36%. Farmer profits dropped 36%.

The year 1985 was the last time the Trussonis sprayed any chemicals on their crops. "Tobacco was dying out and there was talk of exploring organic vegetable production and possibly organic dairy," says Trussoni's son, Arnie. "We were already well on our way to being an organic farm."

People interested in organic were largely attracted to a better price, rather than the philosophy, says the elder Trussoni. "Farmers were being put out of business with lower prices and higher costs, so a different agricultural concept that could sell at a premium price looked attractive." Trussoni knew all too well the age-old problem of pricing. America's farmers didn't have a say in setting prices for their crops, milk and livestock. Trussoni went to work with the National Farmers Organization (NFO) not long after it was founded in 1955 because he wanted to help farmers get a fair shake. The NFO made headlines and earned its activist reputation in 1967 when a group of dairy producers withheld their milk from the market for 15 days to protest food processors' power over setting prices.

> "If we'd had profitable agriculture, you wouldn't see these monstrosity farms that you see today. You'd see family farms all over the countryside generating income." Adolph Trussoni

"The NFO promoted collective bargaining for farmers," says Trussoni. "We advocated a system where farmers could negotiate the price of their production based on their actual costs. Everybody does it, except farmers." The NFO became a producers' union and farmers who signed up named the organization as their bargaining agent. They were an especially powerful force among Wisconsin's dairy farmers.

Living in Tune with Your Turf

Farmers weren't the only activists hanging out in the Driftless Region of Wisconsin during the 1970s and 1980s. Spark Burmaster was the consummate Mr. Fixit who, among other projects, restored and installed Aermotor water pumping windmills and photovoltaic systems. He landed in Vernon County, Wisconsin, purely by accident in 1970 when he gave a hitchhiker a ride. Though he is an electrical engineer, Burmaster is not a linear, by-the-book kind of guy. He trusts serendipity.

Vernon County seemed to him the right place at the right time, so he made it his home.

In 1978, a nuclear plant in Vernon County, at Genoa, Wisconsin, was functioning without an operating license. With hearings before the Nuclear Regulatory Commission to secure a license coming up, Burmaster, with others, took action. They formed the Coulee Region Energy Coalition citizens' group to muster opposition and educate people on energy issues with public events and publications. This marked the start of Burmaster's environmental activism.

An environmentalist at heart, Burmaster also was attracted to the bioregional philosophy of organizing in which local people gather and commit to making a bioregion sustainable.

Burmaster attended a meeting of one of the earliest bioregional networks in the country, based in the Missouri Ozarks. There, he discovered a group of residents and "back-to-the-landers" in Missouri working for sustainable economics, environmental integrity and self reliance. Their philosophy struck a chord with Burmaster, who came up with his own simple definition of bioregionalism: "It's the concept of becoming aware of where you're living and then developing a lifestyle in tune with your turf." Back in Wisconsin, Burmaster shared these ideas with his like-minded neighbors, including George and Jane Siemon and Lila Marmel. They spread the word of bioregionalism to other neighbors in Vernon County, and in the fall of 1983, about 50 people gathered at Esofea Park near Viroqua, Wisconsin, to found the Driftless Bioregional Network. This gathering was followed by nine more, setting the stage for creation of a revolutionary organic farmers' cooperative.

"Scattered throughout the hills and coulees of the Driftless area are various pockets of back-to-the-land types: some isolated individuals and families, some close-knit little tribes, all pursuing parallel muddlings toward self-sufficiency, sustainability and stewardship," wrote Bryce Black, an early bioregional advocate from Arkansaw, Wisconsin. "With 'bioregionalism' as a spiffy new rubric for their existing endeavors, these folks made plans to establish a communications network, a barter system to trade skills and wares, and regular gatherings to share knowledge, sustenance and spirit."

Everything Connects

A crisis of conscience propelled Jerome McGeorge out of a corporate job with Sinclair Oil Company in New York City to the Driftless region of southwest Wisconsin in the mid-1970s. It wasn't a direct route.

"New York City was at its criminal worst," McGeorge recalls of the late 1960s. "The city was dangerous. Black anger was at its height and the city's teachers were striking for fairer pay. Here I was, a privileged white guy putting on a three-piece suit every day and going

Driftless Bioregional Network

The Driftless Bioregional Network is a grass-roots organization of people who are empowering themselves to create a viable, sustainable, environmentally sound and self-reliant way of life within the Driftless Bioregion.

We have formed a network to facilitate the networking of ideas, resources, activities; to share skills and knowledge; to increase localized production, to establish community and bioregional trade, to foster the politics of ecology. We meet to tell our stories, sing and create seasonal rituals that draw from our unique culture. We share the joys of life and stewardship of the unglaciated hills and valleys, striving to walk in balance on the Earth Mother.

The snow-covered land of winter gives way to the warmth that opens the buds on the trees, blossoms the spring wildflowers, initiates the planting season. Summer brings heat, thunderstorms and abundant growth, leading to the bountiful autumn harvest season, with woods afire with the hues of change.

Spark Burmaster, 1981; (left to right): George Siemon, Jack Honeywell and Lila Marmel at the October 1983 Bioregional Gathering in Esofea, Wisconsin →

off to a Rockefeller Center skyscraper. I was anti-war and anti-establishment, a real lefty, and my wife told me I was a flat-out hypocrite. She was right."

McGeorge gave notice and left New York with his family in 1970. They took a camping trip lasting nearly six months and lived in a commune in Tennessee, eventually landing in San Francisco in the early 1970s when the civil rights, anti-war and women's movements coalesced. "It was totally exhilarating: the era of the flower children, the summer of love, the music," he says. McGeorge became a member of the Bay Area Institute, a left wing think tank, and he began working with the Black Panthers, the grape workers and the American Indian Movement. His marriage had ended. Ronald Reagan was governor of California, and McGeorge says he felt "extreme alienation. I couldn't live in that culture anymore."

McGeorge returned to the Midwest in 1974 and became involved in an experimental school in DeKalb, Illinois. From there, he migrated to Vernon County, Wisconsin, with friends who knew the area and wanted to make their lives on the land. They bought 270 acres of farm land, and McGeorge was immediately attracted to the bioregional philosophy espoused by Wisconsin's Driftless Network. "Bioregionalism is a reflection of the classic Greek philosophy of 'monism,' a belief that everything connects," says McGeorge. "It emphasizes the interdependency of all life. Bioregionalism and the organic movement are both expressions of monism."

None of Us Are as Smart as All of Us

That bioregional philosophy suited George and Jane Siemon, a novice farming couple, just fine. They met as students at Colorado State University in the early 1970s: he studied forestry and, later, animal science, and she chose food science and nutrition. Raised in cities, they both gravitated to the country and rented an old farmhouse three miles outside of Fort Collins. They bought a milk cow, baby pigs and chickens and cultivated their first garden. While in college, George worked summers for local farmers, irrigating corn crops and caring for sheep and cattle herds. After graduation, they moved to Iowa City where Jane had a dietetic internship and George worked in big agriculture as a farm hand where he was exposed to chemical farming and animal confinement. The couple attended a natural foods conference in Iowa and met a Minnesota farmer who practiced organic farming in New Ulm. "George visited that farm," Jane recalls, "and it was a turning point." By 1976, they had toured the Midwest looking for farms and chose one in Newton Valley in Vernon County, Wisconsin. They started raising tobacco—a tradition in the valley—and bought 11 Guernsey cows. They grew organic produce, and cultivated their fields with a team of horses. The Siemons were the first young couple with babies in the valley in 15 years.

Veteran farmers in the area were generous with their help and advice, and the young couple realized they had moved into a cooperative, rural culture where sharing ideas and experiences served everyone well. The axiom, "None of us are as smart as all of us," would become a theme for George's life.

In their hearts, the Siemons were "back-to-the-landers," too, rejecting the materialism they saw in modern culture and looking for a simpler, more earth-conscious way to live. They found kindred spirits in Vernon County, and especially at those Driftless Bioregional Network seasonal gatherings created to build friendships and share knowledge through workshops and bull sessions, barn dances and softball games, over long tables of potluck food, and while chanting and drumming to honor the earth, reminiscent of Native American rituals.

Back to the Land . . . For the First Time

The first time Dave Engel met George Siemon, it was at the annual Koch tractor sale in June, 1978, in Richland Center, Wisconsin. "George was looking for horse machinery and it was raining cats and dogs," Engel says. "He was wearing a pair of bib coveralls and a little railroad cap, and he was barefoot." Dave and his wife, Marta, a veterinarian, were city-dwellers who had recently moved to the area from Minneapolis (he was originally from North Dakota and she was from California). They had met in Minneapolis, where Marta was completing vet school at the University of Minnesota, and Dave worked in a dental lab. Upon meeting, they discovered they both wanted to get "back to the land," and, within a few weeks, they had purchased a 160-acre farm in Crawford County in southwest Wisconsin.

"A lot of our talk in those days centered around how to make a living in the country," Dave says. "We eventually decided to milk cows, though that meant finding another farm with a dairy barn."

Vernon and Crawford counties in Wisconsin had a growing alternative community of back-to-the-landers, and some folks started natural food co-ops in Gays Mills, Viola, La Crosse, Richland Center and Viroqua. It was a community of kindred spirits, and the young Engel family fit right in.

When the Engels bought their current farm in April, 1981, they became dairy farmers with a starter herd of 12 cows and two teams of horses. They lived in a 10-by 50-foot trailer, raising three boys from infancy, until they could afford their own farm house ten years later (then they greeted their fourth child, a girl). "It was rustic," Marta says. "Neither one of us came from farm backgrounds." But the instinct to farm was present early. "I can recall getting on the phone when I was a little kid in San Diego," says Marta. "I wanted to know how much hay cost and what it would cost to buy farm land."

Dynamic Tension: a Natural State of Effectiveness

Pam and Tom Saunders, both Minnesotans, graduated from Macalester College in St. Paul, pocketed their degrees and headed to west central Wisconsin to find farm land. "Tom was from farming stock and he knew he wanted to get back to the land," Pam recalls, "but I had no clue about farming. My dad left the small farm he grew up on and never looked back. He didn't speak highly of the lifestyle." The Saunderses headed east to Wisconsin, Pam says, "because land to the south and west in Minnesota was too expensive. The small farming plots across the Mississippi were affordable. People in the Midwestern back-to-the-land movement in the 1970s settled up and down the river, and they tended to come from the Twin Cities, Madison and Chicago."

The Saunderses had lived in a commune in the Twin Cities; they were united by their political and philosophical beliefs and they chose land in Prairie Farm, Wisconsin, for their alternative lifestyle in the mid 1970s. "Back-to-the-land and organic was part political, part idealistic." she says. "I thought dairy farming was an interesting thing to try, but I quickly learned it's not something you just *try*. No, you pour yourself into it."

The Saunderses had four children and became involved in the community: Pam served on the school board and started a community newspaper. "The rural area was a great place to raise kids," she says, "and they learned how to work. All farm kids do. We put up all our own food. We sat around the dinner table and I'd say, 'Look kids, we did all this.' They'd roll their eyes, but now they're happy about it."

Pam distinctly remembers the early 1980s, when several of her farming neighbors were in dire financial straits. "For decades, the message from the USDA was get big or get out," she says. When all that costly debt absorbed by farmers grew, it suffocated many of them.

← Dave Engel, organic dairy farmer, in 1992

The activist in Pam got angry. “We came to the country to build a healthy life and we were being told we had to increase our herd to 80 or 100 cows, when the average was 30. That was respectable. You could make a living. But American farm policy was all about serving big agriculture and big business, not about keeping our rural institutions and family farms vibrant and strong.”

In a state known for its activism and grassroots populism, Pam and Tom found like-minded others and ready venues for their activism. Tom Quinn, another Wisconsin farmer, started the Wisconsin Farm Unity Alliance in the early 1980s, and members of the Wisconsin Farmers Union and the National Farmers Organization supported his efforts. The common goal was fairness to farmers.

In 1981, Switzerland successfully produced the world's first cloned mammals—three mice.

Kitty Pityer, a Milwaukee native who became a back-to-the-lander when she married a struggling dairy farmer, worked with the Wisconsin Farmer Advocacy program aimed at helping distressed farmers by providing mediation with lenders. She was also an officer with Tom Quinn's Farm Unity group. “My husband, Reggie, and I called on Wisconsin farmers to find out what they were facing,” she says. “We worked with them and their lenders to try to hold their farms together.”

Reggie Pityer focused on low prices for farm commodities: “I started to think a conventional farmer is going to have to take what the market offers, but some farmers in the east and the west were starting to do well with organic milk, cheese and produce,” he says. “I wondered, could we organize something around organic? Could we get farmers to believe in it?”

About the same time, Dean and Jan Swenson threw their time and energy behind a small farm advocacy organization, started in Madison, called the Wisconsin Rural Development Center. “Wisconsin was losing thousands of farms every year,” says Dean. “Tom Lamb was inside the University of Wisconsin when he started the organization. We stirred things up. The professors had never come up against anyone who questioned them. We introduced rotational grazing and our work helped start the University's integrated agriculture center.”

This group and others proved in those turbulent days of the 1980s that certain kinds of dynamic tension could be very effective in promoting new ideas and shifting conventional thinking about agriculture.

By 1985, the news on the farm was bleak. “Protests over low farm prices and perceived lack of government concern are almost as old as farming, but they have recently increased in number and volume,” Dennis McCann wrote in *The Milwaukee Journal* Sunday, January 27, 1985 edition. “While the rest of the U.S. economy is humming along, the farm economy continues to stagger along under what many say are the worst conditions since the Great Depression.”

In the depths of the farm crisis, two fundamental questions emerged for the farmers of southwest Wisconsin who were about to start a new cooperative: Can America be fed without poisoning the land, the farmers who raise the food or those who eat it? If the answer is “yes,” how can farmers earn a decent living doing it?

Community * *Jim Wedeberg*

I consider myself really lucky; I was a member of perhaps the last generation to grow up in a rural Wisconsin small dairy farming community. Most every farm, then, supported a family, people helped each other with farm work, and there was a real sense of community. Events revolved around the farm and accommodated the milking schedule.

In 1979, I decided to purchase my immigrant grandfather's farm and I continued to exchange farming labor with long-time neighbors, Merlin and Ona Dregne. Merlin knew the benefits of animal manure, lime, crop rotation and cover cropping. He would say, "It doesn't take long to destroy the soil, but years to bring it back to productivity." At that time both Merlin and I were members of a local cheese co-op, which he also served as a board member. Merlin knew that farming was changing, so he researched and shared his vision to transition the co-op into a specialized cheese plant in order to develop a niche to sustain local family dairy farms. But in 1979 bank deregulation caused interest rates to more than double (8% to 17.5%) in a few years. Combined with low commodity prices, over-leveraged farms and high interest rates, many farmers were plunged into an economic crisis.

As the crisis worsened, I got involved with an ecumenical organization called Churches' Center for Land and People, at that time led by Sister Marion Brown. Our local group worked along with others throughout the Midwest, setting up hotlines for distressed farmers to call, offering advice on financing and networking with farmers. There were a lot of hardships in the early to mid-1980s as the farming community witnessed not only the changing face of agriculture, but the exodus of young people leaving the farming lifestyle and the aging farmer population that remained on the land.

What I remember most about Merlin was his strong conviction about building and maintaining soils. I will never forget one of the last conversations Merlin and I had in the fall of 1987, shortly before his death. We were standing on his lawn looking across the highway at the large 3-room rural school (now converted to a salt storage shed for the county) and the church just beyond the school. Merlin talked of his concern for the future of community institutions—school, church and business—as agriculture changed. He stressed to me the importance of thinking outside the box to find a viable future for small family farms.

Looking back over the years, I realize just how much those words mean to me: the value of family, community, staying connected to the land and treating it with the respect it deserves, not just for us here and now, but for future generations.

{ One of CROPP's seven original organic dairy farmers, Jim Wedeberg serves as dairy pool director. He lives and farms with his wife, Julie, and his sons, John and Jake, on their Wisconsin Organic Valley dairy farm. }

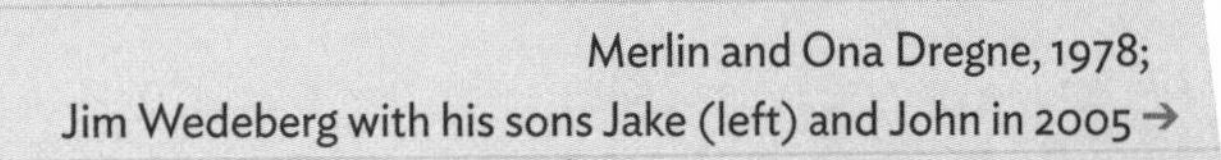

Merlin and Ona Dregne, 1978;
Jim Wedeberg with his sons Jake (left) and John in 2005 →

Bobolinks * *David Kline*

Aldo Leopold writes in "The Farmer as a Conservationist," "It is the individual farmer who must weave the greater part of the rug on which America stands. Shall he weave into it only the sober yarns which warm the feet, or also some of the colors which warm the eye and the heart?"

I grew up with bobolinks. All my life they have nested on our farm and all my farming life I have listened to the males' delightful flight-song from the end of April through early summer.

Why the bobolinks nested on our farm and on very few neighboring farms, I have never fully determined. One reason may be that, although always a dairy farm, we never switched totally from growing red clover to alfalfa, which was where conventional practices were headed in the Midwest.

We stayed with clover because it was more forgiving than the high pH-loving alfalfa and red clover fit a four-year organic rotation better. And for its beauty. Few things are more attractive than a field of blooming red clover being visited by thousands of butterflies and honeybees. Because clover is slower maturing than alfalfa, the bobolinks benefited from the later-cut hay.

For years I wondered why the bobolinks, which arrived on our farm usually around April 28 (the males show up about a week before their mates), delayed nesting until almost June. At the same time the red-winged blackbirds, savannah sparrows, and the eastern meadowlark often had fledged young by the time the bobolinks settled down to nesting, after weeks of chasing each other.

Then I discovered the reason: A decade ago a friend gave me two clumps of big bluestem and I planted them on each side of our mailbox where I see the prairie grasses practically every day. I noticed that by the time our cool season hayfield grasses such as orchard grass and timothy are already heading out in early June, big bluestem had only reached the height where a bobolink could hide its nest. Then I realized that since the bobolinks were native to the prairies, they have it in their DNA to wait for cover before nesting.

That waiting-for-cover conflicts with the bobolink in the eastern United States, where all dairy farmers want dairy-quality hay and that means early cutting, especially on an organic dairy where the prices of off-farm protein are exorbitant. The bobolinks end up being the losers. Only rarely have I seen bobolinks attempt a second nesting. If their nest and eggs or nestlings are destroyed with the mowing machine, bobolinks just give it up.

For that reason, for most of my farming life I agonized over delaying the cutting of one twelve-acre field of hay until July for the sake of the bobolinks. We farm with horses and horses don't need dairy quality hay, I could justify myself in a delayed cutting. But I hated to explain to my neighbors when we had

what we call a perfect hay week in mid-June and we weren't cutting that field of hay.

Sometimes, as Emerson said, a man standing in his own field is unable to see it. Sometimes we need younger and brighter minds to look at the picture in order to get a different and better perspective. I needed my son-in-law's view.

Since we are a grass-based dairy with 50 Jerseys, he suggested we graze the "bobolink field" early but not intensively. Use the bison as a model, he said. We're now beginning our third spring with that method and it has worked way beyond our expectations.

Here is what we do: Beginning as soon as the first males begin singing across the field, he will turn the herd into half of the 12-acre field (the field is divided down the center) for 12-hours, then the other half for 12 hours. Normal practice is to give the cows only about an acre for that time period, which is then grazed fairly short. With that rotation the redwings and sparrows can coexist, but not the bobolinks.

Perhaps three or four days later, depending on the growth rate of the grasses and forbs, the same pattern is repeated. The cows do not disturb the bobolinks because as the grasses begin to reach maturity they tend to follow paths and only meander here and there to graze.

By the end of June the orchard grasses, the dominant forage, are three feet tall and in seed and young bobolinks are flying. While there were tall overripe grasses, there was also a cover of low new growth throughout the field. This year, when I mowed the hay on June 26th, the bobolinks had finished rearing their young and had left the field. The males were already congregating into bachelor flocks several days prior to the mowing of the hay. My son-in-law counted over 50 males in two groups. They were fattening up for their 5000-mile flight to Argentina.

For too long I thought that the songs of our hayfield birds were for their own purposes and our listening pleasure. In other words, the songs were pretty but had no monetary value. Really, money was least of my concerns. Then I found out that the stomata on leaves of the hayfield plants open and close as ordered by the physiology of each species. The stomata open to admit air in order to take in carbon dioxide and trace plant nutrients. Dan Carlson, an innovative plant physiologist, has held and proved with experiments that sound waves from the songs of birds give the plants an assist during the hours of morning mist, for which reason birds broadcast a veritable concerto while the rising sun starts to wake up the rotating planet . . . and the process of photosynthesis begins.

Not only have my eyes and heart been warmed many times by the showy colors and pleasing songs of the bobolinks, when I heard that the birds are an asset to the profitability of our farm I felt like organic cream had just been added to my coffee.

{ Naturalist, writer and semi-retired Organic Valley farmer, David Kline and his wife, daughter and son-in-law farm 150 acres and operate a 50-cow organic dairy near Mt. Hope, Ohio. Kline is the author of three books and editor for *Farming Magazine.* }

2. "We've Got

1985–1988

This Idea"

* The school of hard knocks * Cooperative versus corporat

* The Y in the Road * Positive mission-based business *

George Siemon, the young novice dairy farmer and tobacco grower in Newton Valley, Wisconsin, had found his heart's home place. He and his wife, Jane, were enjoying what he called the "free thrills" of growing their own food.

"The first time I dug my own potatoes, I was thrilled," he says. "I felt like a wealthy person. Jane and I were reclaiming a simpler way of life. We started farming when there were still old timers in coveralls who never missed a milking. They had common sense and old-school values. We were lucky to drink at the well of the past and learn from them."

But the American farm crisis threatened what the Siemons had learned to love. Mixed farm policy messages made it seem as though the rules for successful agriculture were rigged. "Get big or get out," the big boys said, "but *do not* overproduce. If you do, prices will fall and that will be your own damn fault."

For people like George and Jane Siemon and others in the Driftless region of southwest Wisconsin, it was hard to get over one nagging thought: even the most conscientious farmers could not succeed in this contradictory environment.

The School of Hard Knocks

Pam and Tom Saunders, Reggie and Kitty Pityer, Tom Quinn and George and Jane Siemon were among the farmers moved to activism. They joined others who formed a grassroots effort to influence the 1985 Farm Bill, hoping to bring sanity and fairness to the agricultural marketplace. In the process, the "school of hard knocks" would teach them bruising lessons.

These young, back-to-the-land farmers raised on peace marches and the civil rights and feminist movements joined with their neighbors representing generations of Wisconsin farming. Though their politics were often poles apart, they could unite behind this populist effort to preserve the family farm. Average Americans applauded from the sidelines and sent their donations to Farm Aid in a vote of support for the little guys who dared challenge agribusiness policy.

Dateline: Viroqua, Wisconsin. "The seeds of protest are planted in meetings like this—15 farmers around a table in a room above the Viking Inn Restaurant on Main Street talking about a way of life in crisis and what should be done about it." *Milwaukee Journal* reporter Dennis McCann described a meeting in January 1985, when Tom Quinn of the Wisconsin Farm Unity Alliance and Tom Saunders of Prairie Farm urged farmers to "put ourselves back in the process" and show up for a Farm Relief Conference in Wisconsin's state capitol. Other rallies like it were under way in several Midwestern cities.

"In Essentials, Unity"

These modern farm activists had predecessors many decades before them. Cooperatives were born out of grassroots initiatives in Colonial America when farmers joined together to keep their costs low through joint purchases of supplies such as seed, feed, equipment and tools. Some were marketing co-ops created to help farmers get the best prices by combining their production and selling in larger quantities. Other co-ops provided storage, such as grain elevators, or processing: making butter and cheese.

After America's Civil War, the Grange got started to bring farmers together and work for their common economic and political well-being. At its peak in the 20th century, the Grange had one million members who shared a simple and powerful motto: "In essentials, unity; in non-essentials, liberty; in all things, clarity."

The National Farmers Union was another activist organization created to preserve small farm stability and income. Founded in 1902, it created a powerful marketing co-op, supported the Federal Farm Loan Act in the

← One of four farmer meetings in 1988 in the Vernon County Courthouse

← Farmers protested in Washington, D.C., in 1979; a rural Wisconsin farm auction in 1987

Neighbors pitched in to help build each other's barns, 1987. →

early 1900s, promoted rural health co-ops and backed rural electrification. During the 1980s farm crisis, the Farmers Union threw its support behind a court ruling that required the federal Farmers Home Administration (FHA) to disclose options to stricken farm borrowers who were in danger of losing everything.

When the National Farmers Organization (NFO) was born in 1955, it became a fierce bargaining agent for its farmer-members. It negotiated contracts with food processors who bought what its members produced and it engaged in much-publicized strikes called "holding actions" to oppose what they considered unfair pricing. In Wisconsin during the 1980s, the NFO was a respected champion of family farms, and it wasn't afraid to back unconventional ideas.

"We're on Our Own"

Tom Quinn and Tom Saunders, the pair who rallied farmers in Viroqua, were convinced that President Ronald Reagan's farm policies were on a collision course with rural America, and they advocated a different approach. "We proposed a supply management program like tobacco and sugar producers had practiced for years," Pam Saunders recalls. "The farmer has a production allotment, and if he produces over that base, he gets paid less. There is no incentive to overproduce and the allotments keep the price up and stable."

They took their message to anyone willing to listen and especially those who might influence the 1985 Farm Bill. "We put out a tremendous effort, and it was a huge sacrifice," Saunders says. "Our guys attended every meeting they could. They traveled around the state. If they were low on gas, they'd pass the hat to fill the tank and get to the next meeting. Meanwhile, their wives ran the farms and cared for the kids. During that whole fight, our farmers—many who were using organic practices and working their fields with horses—never identified themselves as organic. Our activism was all about saving family farms. When we lost the farm bill fight, the Grateful Dead lyrics came back to us."

> "The future's here, we are it . . . .
> Cause it's all too clear we're on our own."
> *Throwing Stones,* Grateful Dead, 1982

By 1986, the total number of farms in America had declined by almost two-thirds. Shockingly, that implosion occurred in just 50 years. The surviving larger farms dominated food production, and rural communities that were once self-sufficient were steadily declining. It seemed that agricultural policy was intent on reducing the number of farmers in America and promoting large-scale industrial operations.

The unintended consequences of this policy began to surface. With greater use of pesticides and fertilizers during that same 50-year span, the soil was eroding and groundwater was becoming polluted. Americans' health was more closely linked to the safety and quality of their food. It seemed that the country was relying on a modern food production system that was increasingly dangerous and unsustainable.

> "Organizing farmers is expensive and difficult. Frankly, none of us has been very damn good at it. We've got to get organized, but we've been too busy doing chores."
> Tom Saunders

Let's Get Organized

When their efforts to influence the 1985 Farm Bill failed, farmers around Viroqua, Wisconsin, decided they *had* to get organized. "Tom Saunders and Tom Quinn were the primary thought leaders," George Siemon recalls. "They said, 'We can't count on Washington, D.C., to solve our problems. We have to solve our own.'"

Reggie Pityer, a member of Quinn's Wisconsin Farm Unity Alliance, was the first to propose an organic vegetable cooperative. "It was Reggie's brainchild because some of us were growing organic vegetables already," says Siemon, "including a newcomer to Newton Valley who became my neighbor, Richard deWilde."

Though he was trained as an engineer, deWilde soon realized it wasn't a career for him. But farming was. "I had two grandfathers who were avid organic gardeners," deWilde says, "and my mom and I tended our family organic garden. When I left engineering in 1973, I decided it was easier to get started in vegetables than buy a herd of cows. Besides, I was a vegetarian." He leased a small farm south of St. Paul, Minnesota.

DeWilde had been inspired by Rachel Carson's *Silent Spring* and Frances Moore Lappé's *Diet for a Small Planet.* Though there was no established market for organic vegetables yet, Minneapolis and St. Paul had natural foods co-ops that bought deWilde's produce. When the St. Paul farm he leased was destined for development, he relocated to Wisconsin, another early haven for natural and organic foods.

DeWilde was attracted to the forested, rolling hills and winding rivers of the Driftless region of southwest Wisconsin. He attended the founding gathering of the Driftless Bioregional Network in Esofea Park near Viroqua and met Spark Burmaster and George Siemon. It was October 1983. When deWilde mentioned that he was looking for a farm, Siemon spoke up: "There's one for sale next to me."

DeWilde camped on the land in Harmony Valley over July 4, 1984 just to see how it felt. "I remember seeing one car go down the road in two days," he says. "I'd been living in St. Paul in the pathway of an airport with jets overhead. I loved the quiet solitude." In 1985, deWilde produced his first organic vegetable crop on

← Every tobacco plant was touched, appraised, lifted and carried many times; Richard deWilde in 1988

A Wisconsin vegetable farm →

that land. He remembers that area farmers were especially interested in organic.

Indeed, they were. Something had to replace once-profitable tobacco when America's manufacturers turned their backs on Wisconsin's high-value crop.

"There wasn't an organic option for milk, but there were direct marketing options for vegetables. I decided to go with them 100 percent." Richard deWilde

"He's a Tobacco Man"

Brian Rude, a Wisconsin state senator who represented that area of Wisconsin during the mid-1980s, remembers tobacco's demise. "George Nettum was Mr. Tobacco," Rude says. "He was the CEO of the Northern Wisconsin Cooperative Tobacco Pool and they had warehouses in Genoa, Coon Valley, Viroqua and throughout the Kickapoo River valley. (In spite of its name, the cooperative's headquarters was in Viroqua.) George had been promoting tobacco all his life. He believed in the value it brought to producers, but he also saw the handwriting on the wall."

"George Nettum knew the tobacco co-op's days were numbered." Brian Rude, former Wisconsin state senator

Nettum was a national figure, regularly going to Washington to lobby for tobacco farmers. He served on the Vernon County Board for 35 years, and his civic commitments leaned toward environmental and societal issues.

Like George Nettum, Wendell Berry, a tobacco farmer, author, educator and anti-war activist, understood the romance of tobacco. "Burley tobacco, as I first knew it, was produced with an intensity of care and refinement of skills that far exceeded that given to any food crop that I knew about," Berry wrote in his book *Sex, Economy, Freedom & Community* (1992). "It was a handmade crop; between plant bed and warehouse, every plant, every leaf, was looked at, touched, appraised, lifted, and carried many times. The experience of growing up in a community in which virtually everybody was passionately interested in the quality of a local product was, I now see, a rare privilege. As a boy and a young man, I worked with men who were as fiercely insistent on the ways and standards of their discipline as artists—which is what they were .... The accolade 'He's a *tobacco* man!' would be accompanied by a shake of the head to indicate that such surpassing excellence was finally a mystery; there was more to it than met the eye."

Tobacco growing also reflected the cooperative spirit—a philosophy embedded in Wisconsin's agricultural tradition. It was, as Berry says, "a very sociable crop. 'Many hands make light work,' people said, and one of the most attractive customs of our tobacco culture was 'swapping work.'" This sharing was a tradition that bonded the people of rural farm communities when the men exchanged help on each others' farms and the women unveiled their best cooking and prized desserts.

Before America's tobacco manufacturers looked offshore for less expensive sources and public opinion labeled smoking and chewing a moral failing (or at least a health risk), tobacco was gold. Dairy farmers were usually the "tobacco men" in the southwest Wisconsin counties of Vernon, Crawford and Grant. The rolling, forested terrain lent itself to one-, two- and five-acre tobacco plots. They needed tobacco's income to survive.

By law, farmers could only produce a limited amount of this high-value crop for market because they had annual allotments to honor. If they did, the farmers were guaranteed government support. (The allotment and price support system ended in 2005.) Even so, the farmers had long ago organized into tobacco cooperatives to negotiate the best price for their crops, provide heat curing, specialized warehousing, education and a ready source of help. The tobacco co-op also employed large numbers of women to hand-roll cigars and painstakingly grade tobacco, leaf by leaf.

Wisconsin's tobacco growers had consistently lucrative returns for decades at precisely the right time: dairy farmers received their tobacco checks just weeks before their taxes were due.

> "... the tobacco market was the only market on which the farmer was dependably not a victim." Wendell Berry

Wheel of Fortune

"Here was the big question: What were these local farmers going to grow *instead* of tobacco?" Jack Pfitsch says. Patty and Jack Pfitsch were back-to-the-landers in Gays Mills, Wisconsin, about 30 minutes south of Viroqua. She was an author of children's books and he taught high school math. They also grew organic asparagus. "People like us had moved to the area and many were growing organic vegetables, mainly for ourselves," Pfitsch says. "We were interested in clean food and we had already formed food buying clubs and co-ops to find good sources. We knew the demand existed." As it turned out, this area of southwest Wisconsin was right in the center of a geographic "urban wheel," also called the "circle of lights."

"Satellite photos of the upper Midwest showed the concentration of lights going from Chicago north to Milwaukee, west across Wisconsin to Minneapolis/St. Paul, south to the Quad Cities, then east to Chicago," says Pfitsch. "It formed an urban 'wheel' of lights and we were right in the middle. That satellite image showed that our area could become a major clean food source for the urban wheel."

Over café coffee and church potlucks, folks talked, imagined and challenged assumptions. Many of them knew each other and—as the talk continued—the web of proponents grew. They were activist farmers who were fed up with trying to influence agricultural policy in Washington. They were multi-generational farmers who believed their homesteads and farm life were sacred and worth saving. Among them were Driftless Bioregional people who believed in sustainability and stewardship of the land and back-to-the-landers who rejected modernity and sought a purer, better life.

They imagined a cooperative venture based on organic that could be financially sustainable. One that would help farmers stay on the land they loved and produce food free from chemicals. They believed their idea was timely and smart, but they never dreamed it would one day become an agricultural trend-setter.

← (Left to right): Greg Welsh, George Nettum and John Bosshard were early champions of CROPP, and mentors to George Siemon.

One of many potluck meetings of CROPP's founders and families in the 1980s →

"We've Got This Idea"

All that talk coalesced into a series of informal meetings in farmstead living rooms and kitchens in December, 1987. The gathered group included George Siemon, Richard deWilde, Reggie and Kitty Pityer, Spark Burmaster, Arnie Trussoni and Lila Marmel. They called themselves "the Marketing Project Committee," a lackluster label for a diverse group of idealists who would start a revolutionary organic cooperative with farmer-members, products, markets, investors, lofty aspirations and—eventually—profits.

George Siemon called State Senator Brian Rude around Christmas, 1987. "We've got this idea," Siemon told the Senator. "We're trying to put together a new group of organic producers to form a cooperative and I'd like to talk to you." A few days later on a Saturday morning, Siemon stood in Rude's kitchen describing his dream. "I thought it might be a low impact idea, a regional cooperative serving local needs," Rude says. "It was a good idea to get our farmers thinking about alternatives. The fact that it would become a national player never seemed possible."

You Call it CROPP?

The new organization needed a name. Spark Burmaster, the windmill wizard and bioregional advocate, suggested CROPP. The acronym had to be meaningful and pronounceable, Burmaster said, and it stood for Coulee

-3-

Some ideas include Hills & Hollows Organic Co-operative (or Marketing Assn), Coulee Region Hills & Hollows, Good Earth, Bountiful Harvest, Coulee Region Organic Products, Nature's Gifts, Country Harvest, Heartland, Harmony, Upper Mississippi Valley. George would like to include "pool" in the name.

... we need to look at the tobacco pool buildings and ... hear from the attorney with his ... Richard about what job

12/3/1987

-3-

member possibilities are Chuck or Karen Dahl (representing Viroqua), Larry Fitzmaurice (extension), Arnie Kuehn (retired banker), Adrian Hendrickson (banker), Jim Mathias (attorney) and Sister Audrey (religious leader).

There was strong feeling to identify and stand firmly by our "intents." They were identified as: 1) being organic (and distributing an information sheet on why that is important to us), 2) remaining independent (and not accepting grant money or government with strings attached, and not letting outside influence determine our goals), 3) selling at a fair price (and not giving in to the concept that we need to compete on price with commercial products), 4) remaining loyal to local culture (the Farm Marketing and Processing Cooperative is intended to provide farmers in the Tobacco Cooperative with an alternative source of income that can utilize the experience and tradition of small scale, labor intensive, production and cooperative marketing which already exists in the area. The project will start small, but fully intends to develop into a major economic factor in the economic future of the community, and to do so in a way that develops and enhances the sustainable agricultural base -- par. 4 of the project summary), and 5) education.

Our marketing plan for year one -- 1) local market - saturate the local, semi-direct market, 2) contract growing (at least 50% of our acres should be contracted in contracts that are non-binding on our part), and 3)wholesale organic market (direct to stores) - Madison, Milwaukee, Chicago route.

The name and logo are extremely inportant and we need to spend considerable time deciding on them.

Structure of the cooperative: we need to decide if we'll merely rent space from the tobacco pool or use their space, have them send out a newsletter for us and possibly let them handle the accounts.

Farmer/members could be charged a one-time fee of $500 payable out of their season end dividend over three years to avoid excluding an interested grower who can't handle the fee.

A 10% charge on all sales would be in addition to the membership fee.

In addition to memberships, sponsoring shares ($2,000) could be sold to businesses and other interested parties in the community. Membership money and sponsoring share money could be used for capital expenditures and grant money could be used to cover expenses.

We will try to get our hands on a video on organics distributed by Farmers Wholesaler. It is supposed to be very good and informative.

Respectfully submitted,
Kitty Carlson Pityer
Kitty Carlson Pityer

← CROPP's first annual meeting in Coon Valley, Wisconsin, in July 1988

Gary Zimmer presenting at the Midwest Bio Agriculture Seminar in Soldiers Grove, Wisconsin, May 1988 →

Region Organic Produce Pool. It was a mouthful, but it was accurate. Burmaster's friend, George Siemon, emerged early as a natural leader. He could sum up the big picture and drill down to the principles of CROPP's origin with ease:

"When consumers buy food in the grocery store," Siemon explained in 1987, "they have the opportunity to choose what type of food production system they want to support. If they buy conventionally-produced food, it's probably raised in a manner that's harmful to the earth, most likely harmful to agriculture and harmful to people. Organic food represents stewardship of the earth, stewardship of communities, and people who care about personal health."

Siemon decided to take a break from farming in the fall of 1987 and gave himself "permission," as he calls it, to invest more energy in CROPP, at least until his heifers were old enough to milk. Like it or not, the decision would take over his life.

In 1987, more than 89 million acres of U.S. land were treated with neuro-toxic pesticides.

"Organic: a philosophy and system of production that mirrors the natural laws of living organisms with emphasis on the interdependence of all life." Original CROPP by-laws, 1988

Dream On

Big Ag was not shaking in its collective boots when CROPP's volunteer steering committee invited farmers to four "growers meetings" on January 18, 26, February 2 and March 2, 1988. The plan was to introduce CROPP and determine if there was enough local support to make a go of this regional organic produce pool. The steering committee included George Siemon, Spark Burmaster, Gunars Petersons, Guy Wolf, Reggie and Kitty Pityer.

"What do those pipsqueak farmers with fewer than 100 acres think they're doing?" the boys from Big Ag probably snorted when they heard about the venture. "Start a new venture in the midst of an ag crisis? Do it without proven herbicides, pesticides and chemical fertilizers? Dream on, boys."

"Once we identified the model, we hoped people would be able to see themselves in it." George Siemon

CROPP got its start *because* of the 1980s farm crisis, not in spite of it. Yes, its founders were idealists and dreamers, but they were practical, too. If they were to succeed, they had to create a viable economic alternative for family farmers. This new model must prove itself or common sense-driven farmers would never buy it.

There would be no dissing conventional ag, either. CROPP would simply propose another way to farm. "We were ridiculed about our ideas of organic," says Siemon. "In the beginning there was a lot of tension. We had to be cautious about criticizing conventional agriculture. We discussed alternative choices and avoided conflict so that we could be part of our rural communities. Our success was dependent on a good relationship with conventional agriculture, both its infrastructure and also as a future source of organic farmers."

CROPP's invitation poster to the four grower meetings beginning in January, 1988, was thumbtacked to bulletin boards in churches, coffee shops, feed stores and implement dealers. It said simply this:

"A number of local farmers are marketing produce as cash crops. They have found a huge and expanding demand for quality, organic, fresh and processed produce. To meet the expanding demand, more growers are needed along with an efficient marketing system. Therefore, a coulee region organic produce pool is being organized.

"The cooperative's goal is to be an efficient, economically and environmentally sound, self-sustaining marketing system, and thereby enhance enduring rural development.

"The organic product is grown without harm to the environment or groundwater, is nutritious, residue-free, and returns a higher price to the grower.

"The initial emphasis is on vegetables and fruits, with future expansion into other locally-grown products such as grains, animal and dairy products."

This language—first used to describe what some thought was revolutionary and others called folly—has changed little. Twenty-five years later, an older and wiser George Siemon describes the power of CROPP's positive, mission-based business. "A business is defined by its mission," he says. "We had a hopeful vision to establish a business that embraced organic farming that would enable family farm agriculture to be sustainable. This sustainability starts with care of the earth, but as a business mission, it also has a larger focus on economic sustainability—both for CROPP and the family farmers we serve. CROPP has been blessed with a clear, positive mission that has guided us from the beginning."

Power in Diversity

The farmers at CROPP's first meeting on January 18, 1988, were probably shaking in their Sorels. Wisconsin was locked in a relentless winter deep-freeze. Even so, more than 180 people walked into the Viroqua County Courthouse that evening, kicked the snow off their boots and settled into courtroom chairs. Not counting spouses and family, it was about 160 more people than the organizers expected. "We were dumbfounded and overwhelmed," Siemon recalls. "We really didn't know if we'd even attract 15 people."

Word of mouth, local newspaper articles and the invitation posters had brought together a surprisingly diverse group.

"I read about the first growers meeting in the Vernon County newspaper and I knew nothing of the idea," Jim Wedeberg, a fourth generation dairy farmer, says. "I recognized George Siemon's name and I told my wife, Julie, that we should go. We took a couple friends, including Jack and Patty Pfitsch. The courthouse room was full, and I was surprised by the mix of people. There were the back-to-the-land folks from the alternative community, traditional farmers who had lived in Vernon County all their lives, and a lot of Amish families."

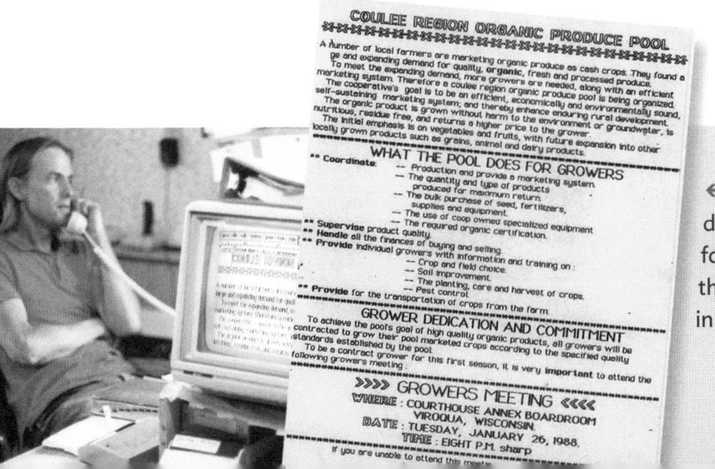

COULEE REGION ORGANIC PRODUCE POOL

A number of local farmers are marketing organic produce as cash crops. They found a ge and expanding demand for quality, **organic**, fresh and processed produce.
To meet the expanding demand, more growers are needed, along with an efficient marketing system. Therefore a coulee region organic produce pool is being organized.
The cooperative's goal is to be an efficient, economically and environmentally sound, self-sustaining marketing system; and thereby enhance enduring rural development.
The organic product is grown without harm to the environment or groundwater, is nutritious, residue free, and returns a higher price to the grower.
The initial emphasis is on vegetables and fruits, with future expansion into other locally grown products such as grains, animal and dairy products.

WHAT THE POOL DOES FOR GROWERS

** **Coordinate**: -- Production and provide a marketing system.
-- The quantity and type of products produced for maximum return.
-- The bulk purchase of seed, fertilizers, supplies and equipment.
-- The use of coop owned specialized equipment
-- The required organic certification.
** **Supervise** product quality.
** **Handle** all the finances of buying and selling.
** **Provide** individual growers with information and training on :
-- Crop and field choice.
-- Soil improvement.
-- The planting, care and harvest of crops.
-- Pest control.
** **Provide** for the transportation of crops from the farm.

GROWER DEDICATION AND COMMITMENT

To achieve the pools goal of high quality organic products, all growers will be contracted to grow their pool marketed crops according to the specified quality standards established by the pool.
To be a contract grower for this first season, it is very **important** to attend the following growers meeting :

>>>> GROWERS MEETING <<<<
WHERE : COURTHOUSE ANNEX BOARDROOM
VIROQUA, WISCONSIN.
DATE : TUESDAY, JANUARY 26, 1988.
TIME : EIGHT P.M. sharp
If you are unable to attend this meetin

← George Siemon drumming up support for CROPP; invitation to the first farmers meeting in January 1988

A farmers meeting at the Vernon County courthouse →

"We have proven that dedication to a common mission will transform diversity into foundational strength." George Siemon

Jerome McGeorge, who ditched corporate life and wound up in Vernon County's alternative community, would soon become CROPP's chief financial officer. He had sought out the Amish to learn about organic farming practices and to understand what McGeorge calls "the spirituality in farming" and how it expresses itself in Amish daily work. McGeorge believes that the broad diversity exhibited at those growers meetings—and especially the participation of Amish farmers—was one of CROPP's earliest strengths.

It was powerful, says Jack Pfitsch, because people saw others who were very different from themselves, but they were all still attracted to the CROPP vision. "I only remember two times when the Amish have joined with 'the English,' as they call us, to unite behind a cause—this meeting, and protesting low-flying military planes over the Driftless region." The Amish also stood up for homeschooling and against outlawing outhouses, and they won both times.

"Probably the greatest difference between Amish and conventional agriculture is the community life or support we have." David Kline, Amish organic dairy farmer, Fredericksburg, Ohio

Confronting "Chaotic Consensus"

Guy Wolf, a North Dakota farm boy and rural organizer, had been drawn to the activist farmers who envisioned CROPP. He was also the meeting facilitator at that first growers gathering. "We learned a lot from the people who showed up," he says, "not only who they were, but what they were struggling with. Crop prices were awful, and dairy farmers were hit hard, too. The young people who wanted to have a future in farming didn't have much hope. People were interested in organic, but many were just looking for a way to survive. None of us made any judgments about who was who or why they came. We were all in this together."

In those four meetings from January through March, 1988, George Siemon described what farmer-members could expect from CROPP and why a pool was a smart business model. Up front, Siemon said that CROPP's original steering committee didn't have a corner on the best ideas. He told the assembled farmers that CROPP's direction should reflect the vision of the farmers attending these meetings. It was probably Siemon's first acknowledgement of "chaotic consensus"—a style of decision-making that would characterize CROPP's origin and evolution. "In our decision-making process, we have always strived to have broad buy-in," Siemon said years after CROPP's founding. "In the midst of one of our chaotic conversations, folks might feel that we're not getting anywhere, but suddenly a consensus forms from the chaos." It did in the winter of 1988, when consensus mattered most.

"There was a lot of brainstorming, especially in CROPP's first year," Guy Wolf recalls, "and many late evenings involving some people whose politics were somewhere near anarchy. There was a lot of talk about what CROPP might look like and what it might be. I remember that all ideas were welcomed." That's why meetings often lasted late into the night. Finding consensus in chaos could be agonizingly slow, but the results proved powerful.

A Cooperative of Organic Pools

CROPP's steering committee told growers that the Cooperative would provide refrigeration for their produce,

What's the Better Alternative?

Though he had no formal business training, George Siemon, the youngest of four boys, grew up in a business family where talk of profit and loss infused dinner table conversation at their home near West Palm Beach, Florida. Siemon's maternal grandfather, James Waugh, owned a local office supply business that he built to seven stores and a distribution center. "When they had frustrations about the business, they always said, 'Well, it's really all about our 100 employees and their families,'" Siemon recalls. Business meant family. Business meant community. It also meant forever. "They were all about passing the business on to the next generation, instead of selling out. I realized that family businesses are very much related to family farms."

Even when Office Depot landed in south Florida, the family hung on through the decimation of downtown areas, construction of suburban shopping malls and their own financial failure. "Watching my family go through hell with business bankruptcy was a real learning experience," says Siemon. "I'd never been a very profit-oriented person, but I realized that a profitable company allows a mission to flourish. You may start out with noble values, but if you don't have a viable business plan, the values don't carry the water."

In fact, no one was more surprised than Siemon that he became CROPP's first (and only, to date) "CEIEIO." As a kid, he loved nature and the outdoors. With three older brothers, Siemon retreated to nature. Seeing a section of the Everglades bulldozed so Minute Maid could start a large-scale orange grove angered him. "I lived in a place where things were getting destroyed that would never come back," Siemon told journalist Burt Berlowe. "I had a concern that the next generation would never experience the things that were dear to me."

In his youth, Siemon spent several summers on farms owned by his extended family in Iowa and Alabama. He raised a pet raccoon, kept his own small brood of chickens and took to bird watching. Not surprisingly, Siemon earned a degree in animal science at Colorado State University before he and his wife, Jane, became diversified dairy farmers in southwest Wisconsin.

"Jane was a wonderful farmer. She wanted nothing more than to stay on the farm," says Siemon's mother, Margaret. "But George had ideas and he decided to do something about it. Nobody thought they would amount to anything, but it was a mystical time. The result was pretty amazing."

"I had to face the reality of what my dreams meant," Siemon told journalist Berlowe in 2011. "Good ideas usually come with a workload. I believe the ultimate thing is to prove that business can be done differently . . . I've never been one to fight the status quo. I just ask, 'What's the better alternative?'"

George Siemon (center) with horse farmers Steven Adams (left) and Bill Kopplin in the 1980s

← (Left to right): Guy Wolf, George Siemon and Tom Quinn in March 1988

Organic farming education at the Western Wisconsin Technical College in La Crosse, Wisconsin, April 1988 →

storage, marketing of their organic crops, bulk purchase of supplies and much-needed education. CROPP might even get into processing. A produce pool was the ideal vehicle for CROPP, Siemon said, because farmers joining the pool could cut costs by purchasing in bulk and win a better price for their production if they marketed and sold their harvests together. In addition, the farmer-members would control the pool's function and future. In fact, CROPP could eventually become a "cooperative of pools" organized around a variety of organic products—vegetables, dairy, meat, you name it.

> **"The idea of pools originated with tobacco. Farmers could pool their production to share the benefit of a higher price and ride out the down times better together."** Jim Wedeberg

Even though many of CROPP's first produce pool members were inexperienced organic growers, some with poorly-prepared soil, members of CROPP could learn from more knowledgeable farmers, including Richard deWilde—a successful organic producer with credibility—and agricultural experts at the area's colleges. This educational component was big from the beginning, making CROPP a "learning organization" years before the term became a popular business buzzword.

CROPP's young leaders were excited by the new venture, but they also faced a steep learning curve. They had little experience in running a cooperative business and they had farms to run and chores to do. But they were willing to do their best, find the lessons in their mistakes and keep getting better. Learning from the "school of hard knocks" was a persistent theme in CROPP's early years.

Though organic certification was still in its infancy both locally and nationally, Richard deWilde, as the resident expert organic grower, told his fellow farmers that consumers were increasingly concerned about the healthfulness of food. He said that reliable certification would eventually build support for organic. In the meantime, he said, the "surest market" was urban areas where consumers were open to organic products and accustomed to paying more for quality. DeWilde said he had successfully sold his organic produce in Minneapolis, Madison and Milwaukee. Perhaps Chicago would be the next stop for CROPP?

"We must offer top quality if we expect consumers to buy from us," deWilde said. That would not come easily, he warned.

At first, some CROPP growers would be producing organic vegetables on soil that wasn't properly prepared with compost or a "cover crop." With effort, however, the soils would improve. Look for lower than normal yields in the first year and a high number of crop failures, deWilde cautioned. There was more to organic farming than just eliminating chemicals.

DeWilde asked for a show of hands at the third growers meeting: how many farmers have "certifiable soil" that would pass a test now? About half raised their hands.

Rushing Forward

"Some people said we shouldn't rush forward in 1988—we should wait," Siemon recalls. "But others said, 'Let's go.' We had to either get CROPP on the road or lose our chance. We rushed forward without having all of it figured out. We were a bunch of do-gooders with a concept that was complex. We were taking on too much by creating a whole new working model of a co-op at the same time we brought produce and organic growing skills to a whole new set of growers. Had we waited another year, we would not have had that foolish enthusiasm that carried us through the first year."

> "We were brave and foolish." George Siemon

"There were so many questions," Guy Wolf says. "All we could say was, 'We don't know. We'll work on it.'" Basic questions, such as what will we do with the bruised tomatoes or nicked zucchini? How do we create a market for seconds? "We'll get back to you," CROPP's overloaded steering committee answered.

To that question of seconds, an Amish farmer offered his recipe for organic vegetable juice. Siemon's enterprising wife, Jane, began baking and selling zucchini bread, following her mother Lillian's recipe, made from bruised seconds.

> "When we founded CROPP, a lot of us didn't like co-ops. We didn't want to start another one just like the rest of them." Wayne Peters, organic dairy farmer and original CROPP dairy pool member

Cooperative Versus Corporate

Historically, cooperation in agriculture made grassroots sense as farmers banded together to market their crops, advocate for better pricing and support small, family farms with lower-cost loans, ag education and a voice in Washington. Aaron Sapiro, a lawyer who grew up in poverty in Oakland, California, in the 1890s, was a major farm cooperative leader during the 1920s. He urged farmers to join together in cooperatives, market their own crops, eliminate the middlemen and dramatically increase their farm profits. Sapiro spent much of his time organizing co-ops in California, especially among fruit growers. The *New York Times* called him "the leader of one of the greatest agricultural movements of modern times." Widely known as the "California plan,"

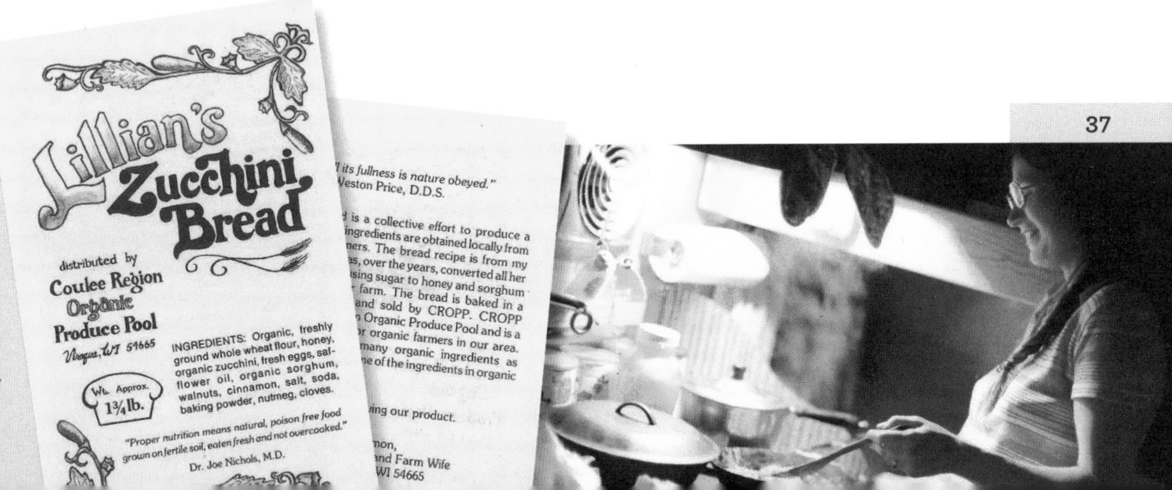

An ace in the kitchen with a degree in nutrition, Jane Siemon made and sold organic zucchini bread using her mother's recipe in 1988. →

Sapiro's model wasn't lost on farmers thousands of miles away in Wisconsin.

"Sapiro was a real go-getter," George Siemon says. "He'd get off the train somewhere in the U.S., give a two-hour speech, hand out information and he'd leave a new co-op in his wake. Then he'd go to the next town. Viroqua, Wisconsin, was one of his stops and George Nettum, Mr. Tobacco, heard him speak." As it turned out, Nettum's Northern Wisconsin Tobacco Pool was a Sapiro co-op and the last of its kind in the country.

"Because he was a lawyer, Sapiro believed he could enforce farmer loyalty with a contract. His system had no soul; it was just business." Spark Burmaster

Wisconsin's first cooperative dairy pool was formed by Anne Pickett in 1841, before Wisconsin was even a state.

But the gulf between Saprio's dream and the reality of running a successful cooperative was often too great. How many farmers had the skills—or time—to start and run an organization, while still farming? Very few, it turned out. These inexperienced co-op founders promised farmer-members that if they invested part of their income in the co-op, they'd be paid back—at a premium—with future profits. But that didn't always happen. Sapiro had counseled cooperative leaders to pay farmers all they could, make few promises for the future and avoid focusing on a primary goal of making money for the enterprise. If a co-op needed more operating income, it should start a related side business, rather than pull funds from the farmer-members' fair share. In some cases, cooperatives seemed more concerned with the profit of the co-op than with the needs of the very farmers they were created to serve. Many American farm cooperatives failed or disappointed their members.

If a cooperative was such a dicey model, why not incorporate CROPP, a few people asked? Like the companies on Wall Street? The answer boiled down to the ultimate goal: corporations were in business to maximize the value of the business. CROPP's goal was to serve family farm members fairly for generations. If that difference between CROPP and a big corporation wasn't clear enough, farmers had only to remember Wall Streeter Gordon Gekko from the movie *Wall Street,* released in 1987, who declared, without apology, "Greed is good."

"A lot of us were unhappy with cooperatives as we knew them," says Siemon."We decided to create CROPP the right way or we weren't going to do it at all. Even if it meant failure."

In the midst of heated debate during one of CROPP's early meetings in 1988, Siemon recalls one farmer, Tom Forseth, standing up and making a case for creating a new kind of cooperative, one with good character. "What would a regular co-op do?" Forseth asked the group. "I say, we do just the opposite!"

"We looked into bylaws for CROPP and some of the most democratic we saw in agriculture were the tobacco pools. Their one farm, one vote rule impressed us." Guy Wolf

An "Auspicious" Start

"The farmers who actually jumped in and decided to grow organic vegetables that first year took a big risk," says Guy Wolf. "I had only about a quarter acre, but I

← Tobacco-growing needs many hands. Neighbors help out on tobacco farm, Vernon County, Wisconsin, 1978.

↑ Richard deWilde's organic vegetable farm in Harmony Valley, 1997; Carol Anne Kemen and Jerome McGeorge plant 18,000 onions in May 1990; (inset) Dave Engel; George Siemon (far left) meeting with the Northern Wisconsin Cooperative Tobacco Pool Board of Directors; a meeting of the CROPP board in 1988 (clockwise, top center): Mike Breckel, Greg Welsh, Reggie Pityer, George Siemon, Richard deWilde, Jay Harris and Gunars Petersons

was one of them." Adding to the uncertainty was this: The year 1988 would bring one of Wisconsin's longest droughts in history.

Things happened fast after those four growers meetings in early 1988. At first, 50 people committed 50 acres of organic vegetable production, but it would soon increase to 60 growers and 75 acres. Their goal for the year was 100 acres and no more. CROPP's first farmer-members each paid a one-time registration fee of $25 to join CROPP, an annual grower fee of $50 to cover soil tests, organic certification and education services, and a "front fee" of $100 per acre per year. This "front fee" paid for essentials including seeds, organic fertilizer, pest control and ingredients used to improve the soil. Farmer-members agreed to observe CROPP's quality and production standards and only sell their committed crops through CROPP.

CROPP created its first board of directors—all volunteers and led by Mike Breckel, an organic farmer, food co-op founder and Peace Corps volunteer. They filed their incorporation papers with the State of Wisconsin on March 10, 1988. Jerome McGeorge, CROPP's financial officer and part-time astrologer, remembers the Sunday before the filing when he, George Siemon and Spark Burmaster gathered to review the final papers. McGeorge had developed an astrological chart entitled "CROPP begins 6 March 1988 Sunday," crowded with symbols and annotations (plus a few coffee stains) that only a trained astrologer could decipher. The chart's findings were auspicious in every way, McGeorge says. "It was extraordinarily idealistic, full of promise and grounded in the earth," he says. "Because the earth element was dominant, it was ideal for an organic cooperative." Having studied and practiced astrology for 20 years, McGeorge says the chart "deeply impressed" him. "The founders of CROPP were inspired," he says, "and they were ready to move." Watching its evolution years later, some insiders would say that CROPP was born under "a lucky star."

> "Gunars Petersons suggested passing the hat to get the pool off to a good start. The donations amounted to $128.87." CROPP meeting minutes by Kitty Pityer, January 18, 1988

Inspired and lucky, perhaps, but CROPP was woefully short on cash. Even the most optimistic projections for organic produce production made its financial projections in that first year look anemic. CROPP had the promise of a modest start-up grant of $5,000 from Tom Quinn and the Wisconsin Farm Unity Alliance, but, at best, that would only pay for phone service, office supplies, copying, a little accounting and legal help and small stipends to cover organizing. "No salary" was the first line item on CROPP's start-up budget. There would be plenty of talk about fundraising as CROPP got on its feet. Grants from government, foundations and corporations were on the table, and so was the sale of stock.

> "Looking at the original minutes in 1988, I was struck by how scared we were. It was obvious that vegetables alone were not going to do it for CROPP." Dave Engel, dairy farmer

"Mr. Tobacco," George Nettum, had thrown his support behind CROPP, too. Nettum's Northern Wisconsin Cooperative Tobacco Pool had unused warehouse space. (Nettum had become an important mentor to George Siemon, sparking him to get involved from the very beginning, even before the formation of CROPP.) Could CROPP convert some of it to store its organic produce? Maybe the tobacco co-op could take CROPP under

-2-

In Kentucky, a cooperative distributes to the local poor and accepts food stamps as well as working with church organizations and legal aid. We could look into something similar to this.

George suggested an organic V8 juice. Processing is on the back-burner right now until we have funds to actively research.

Gunars pointed out that we need to focus more and prioritize our projects. We need to develop board committees to help us avoid going in too many directions. A zero-based budget is practical for our group.

There are many choices to fund this project in addition to grants, such as low interest loans, preferred stock. WFUA did not vote on the $5,000 grant at its January board meeting, but it is scheduled to be voted upon on Friday, January 22. The following tentative budget for the $5,000 was discussed.

No salary

Office		200 supplies 200 postage	
	Total		400
Copying			300
Phone		500 back bills 1,000 future bills	
	Total		1,500
Travel			1,000
Perdiem		$25/day 24 days	600
Other		300 legal 200 market survey 100 accountant 600 misc.	
	Total		1,200
GRAND TOTAL			$5,000

Respectfully submitted,

Kitty Carlson Pityer

Kitty Carlson Pityer
(Minutes typed from notes taken by Sparky Burmaster)

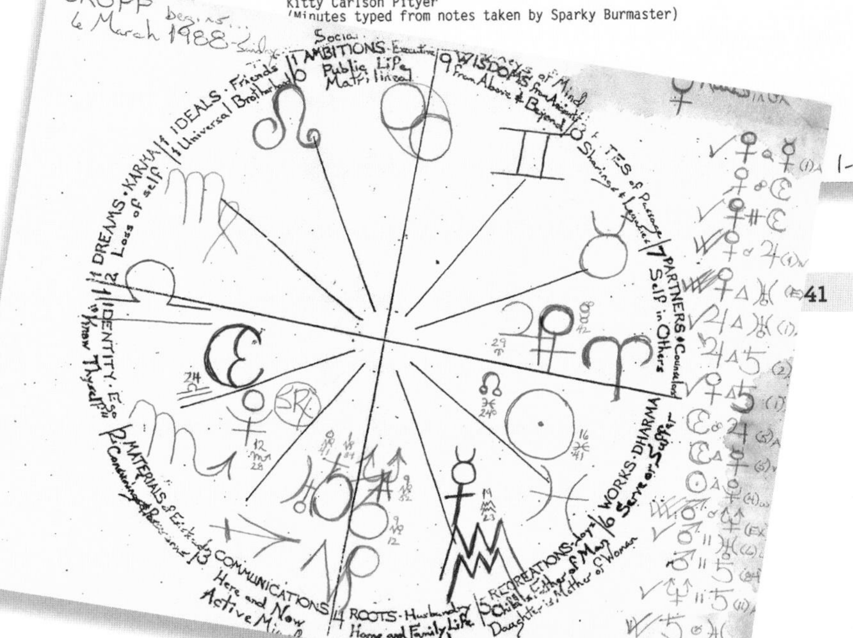

its wing, and Senator Rude could author legislation intended to finance the storage conversion to the tune of $50,000? The proposed legislation would ultimately pass, but be vetoed by the Governor in May of 1988. In retrospect, it may have been for the best. CROPP retained its independence.

Nix Elbows and Ass

Just one week after their incorporation papers were filed in March, 1988, George Siemon brought a proposal to CROPP's board of directors that would transform the future of the organization and fuel financial and geographic growth that no one dared imagine.

He described formation of a CROPP organic dairy pool, with a starting goal of 20,000 pounds of milk produced every other day. CROPP would hire creameries to process the milk and make organic cheese, too. The mighty National Farmers Organization of Wisconsin was even willing to be a partner in the venture.

CROPP's leaders were vegetable growers and their reaction to the new idea was a mix of curiosity and ambivalence.

> "If we're doing this with organic vegetables, why can't we do it with milk?" Dave Engel

This idea of an organic dairy pool actually had its start the previous Christmas, in 1987. Siemon told his friend Ray Hass, a vocal champion of organic, about CROPP and its formation. He hoped Hass would sign on as a vegetable grower. Hass's response to Siemon was cryptic. "George," he grinned, "I'm not into elbows and ass, but if you ever want to do something with dairy, I'm in!" Bending over and picking produce for hours in the hot sun was not Hass' idea of satisfying farming. "Ray planted the seed," Siemon says. "We hadn't even thought of dairy."

Later on, when Siemon described their dairy pool plan to Adolph Trussoni, a key leader of the National Farmers Organization, he said CROPP's innovative pricing approach was, in Trussoni's words, "deserving of support."

> "The question at our board meeting was, 'Are we interested in this dairy project for CROPP?' The answer was 'Yes, I guess, if you think it would work.'" Jack Pfitsch

"At first, the motivation for organic production was not as much organic as it was price," Trussoni says. "Dairy farmers were gradually being put out of business with low prices and higher costs. The farmers who started CROPP decided to focus on organic production and adopt a totally different pricing method based on what the farmer actually needed to generate a reasonable income. I'd been trying to sell collective bargaining to farmers who had absolutely no input into the pricing structure of their products for years. CROPP's approach reflected the way business should be done, and this was the first group who took the bull by the horns and decided to make it work."

Trussoni's son, Arnie, would become an early and ardent supporter of CROPP because he wanted desperately to continue the family farm. Years later, young Trussoni would become CROPP's board president.

> "The NFO always pushed for higher, sustainable prices for farmers." Wayne Peters

The Y in the Road

From the beginning, the founders of CROPP believed that organic farming must yield economic sustainability for its farm family members. Other cooperatives had proven that "what's best for the co-op" was not necessarily best for members. CROPP would take a revolutionary approach that it called the "Y in the Road," starting with the new organic dairy pool. It was revolutionary because it reversed a standard cooperative practice. CROPP farmer-members would be paid *first* based on a fair target price and *then* left over income would be spent on the cooperative's operating expenses. When the NFO became CROPP's financial partner and sponsor in the early days, CROPP's dairy funds would go through the NFO to pay the farmers first.

Choosing the target price for organic milk was equally revolutionary. Rather than basing the price on what the conventional market dictated, it was based on what the organic family farm actually needed to run a sustainable business. "When the CROPP dairy pool started, it was with very strong roots in collective bargaining," George Siemon recalls. "There was frustration with the existing market system, whether cooperative or not. The organic milk market was seen as an opportunity to practice collective bargaining and to seek to establish a stable pay price. We decided to take a firm stand on pricing for organic milk and to never sell to any customer below our established target price. This didn't mean that we ignored the market or competition, but the primary consideration was the economic viability of organic farming."

In 1988, farmers received only $0.88 of the $2.30 per gallon price of milk.

Was this pie in the sky thinking? Twenty-five years of never selling a farmer's dairy production below the established target price would prove that this bold policy could work. It would depend upon carefully thought-out principles, established in 1988, to choose the rough target price, and these principles are still working today.

It took some time before the leadership of NFO's dairy division would fully support the milk pool partnership idea with CROPP—but they did fully support the partnership in time, an enormous key to CROPP's success. "At first, they were reluctant because CROPP looked like a fly-by-night organization," Adolph Trussoni said. "Some of our leaders didn't see any potential, though some of us did." Trussoni had the credibility and persuasive powers to eventually make the sale.

"Adolph stood by us," Siemon remembers, "and Mark Michel was the initial driving force behind our dairy program. Mark was an NFO field rep who really knew the farmers. He was in my house when he told me about this guy, Wayne Peters. 'You really need Wayne,' Mark said. 'He's got a big hunk of milk.' We got Wayne on the phone and I had to hold it a few feet from my ear. Wayne was giving me a lecture on dairy pools and he shouted, 'I'll start my own program if you guys don't make it happen.'"

"If you want a sustainable pay price, you must have sustainable production."
Wayne Peters

Siemon hung up, shaking his head. "Maybe we don't really want that guy in the pool," Siemon said to Michel. "No, George," Michel assured him, "you do." (Peters would eventually serve as CROPP's board president.)

Michel proceeded to visit Wayne Peters and many other farmers in the area and talk up the fledgling CROPP dairy pool. Wayne Peters and Ray Hass were big producers and best buddies. They'd shop for farm machinery in Iowa and hit all the coffee and donut shops on their route. They were NFO members and they knew and respected Adolph Trussoni. "Ray and I were organic farmers and all we were looking for was a decent price for our milk," Peters says. "We never had the idea that someday CROPP would be a national cooperative." Together, Peters and Hass would produce about half the total milk for CROPP's dairy pool in that first year.

Dean and Jan Swenson were among the first nine family farms to join the CROPP dairy pool. "It was the first time I really felt good about a co-op," says Dean, "because the farmers were making the decisions. When CROPP decided to set its own price, I thought that was a real beginning. Up to that time, farmers took whatever price they were given."

"Most of us didn't know what the O word was." Jim Wedeberg

"George gathered about a dozen people who were interested in organic dairy, because he was the only one on the CROPP steering committee with any dairy experience," says Jim Wedeberg. "There weren't any organic dairy standards established yet, but Julie and I had largely stopped using chemicals on our farm in 1980. I was looking for another market for my milk because of rBGH." This controversial artificial growth hormone—recombinant bovine growth hormone—was created to boost dairy cow production and government approval was imminent. It would be another factor that mega-dairies would use to kill off small, conventional co-ops, and Jim Wedeberg had seen enough of that already.

CROPP's first pick-up of organic milk from its original farmer-members occurred on July 12, 1988, and on July 13, CROPP's organic milk was used to make America's first commercial, certified organic cheese at Torkelson's Springdale Creamery, Inc., in Readstown, Wisconsin.

"I remember reading the minutes of CROPP and noticing a statement not long after the first milk pick-up," says Dave Engel. "'It looks like we're going to make $2,700 a month on this dairy project,' it said. 'We should be able to make ends meet shouldn't we?'"

Little did they know then, it was a mere drop in the milk bucket. CROPP's dairy pool members would eventually produce nearly 90 percent of the cooperative's annual revenues to the tune of nearly $860 million in 2012.

Treat the land right and it will take care of you.
ORGANIC VALLEY
FAMILY OF FARMS
Organic Valley Farmer/Owner
Ray Hass ~ Viroqua, WI

← Organic dairy farmers and buddies Ray Hass (left) and Wayne Peters produced almost half of CROPP's dairy pool milk in the first year.

↑ Images of organic farming in southwestern Wisconsin →

Organic: Beyond the Farm * *Frederick Kirschenmann*

During the past twenty years, organic farmers concentrated almost all our energy solely on our own farms. We improved our operations and developed markets that had the potential to keep the farm going.

Now we're faced with daunting new challenges: the end of cheap energy, depleting fresh water resources, and more unstable climates. Coping with these new realities will require a new paradigm utilizing adaptive management and cooperative relationships with others in the industry.

Fortunately, organic farmers are already on a path that enables us to draw on wisdom from the past as well as new science. In Grass Range, Montana, Tom Elliot owns and manages a large cattle ranch. Several years ago, Tom noticed that since his cattle didn't eat leafy spurge, it was starting to take over his rangeland. He called chemical advisors, developed an eradication program, and helicopters sprayed the spurge. The price tag was $25,000.

The next spring the same thing happened. Tom called his advisors, who sprayed his range for $25,000.

The third spring the spurge still came back. What's more, his pine trees were dying and he had more sick cattle. Tom decided chemical sprays were not the answer.

Then Tom heard that sheep loved spurge. He called a neighbor with a large herd of sheep and inadequate grazing land. Together they drew up a lease agreement that allowed the neighbor to graze his sheep with Tom's cattle. The spurge provided excellent pasture for the sheep; they kept it in check without depleting the pasture for the cattle. And Tom received a handsome income from the lease. The new culture of cooperation brought success to both Tom and his neighbor.

Resilience thinkers from the new professional societies, the Resilience Alliance and the Ecological Economics Society, remind us that economic, social and natural systems are always changing due to circumstances largely beyond anyone's control. Diversified farms might not be as efficient in the short term as monocultures, but they will likely be more resilient in the long term. If energy costs and unstable climates are devastating for one crop or one kind of livestock, it may not be devastating for every crop or livestock. During the drought of 2012, the rye crop on my North Dakota farm was great (rye had matured before the most severe drought, as did the flax). The native prairie's perennial root system survived the drought quite well, providing adequate grazing for our livestock.

By working together in cooperative arrangements—like Organic Valley does—we may be able to reinvent organic agriculture to serve the needs of farmers, consumers and the planet in our new world.

{ Frederick Kirschenmann is a Distinguished Fellow at the Leopold Center for Sustainable Agriculture at Iowa State University, and President of the Stone Barns Center for Food and Agriculture in Picontico Hills, New York. Kirschenmann also runs his family's organic farm in North Dakota. }

From Rag-Tag to Big Time * *Frances Moore Lappé*

Selected from *Hope's Edge* by Frances Moore Lappé and Anna Lappé, Penguin Putnam, 2002

When we heard that we'd be spending our last day in Madison at the Willy Street Food Co-op, I was dubious. My memory of Willy Street was a tiny storefront; I couldn't imagine enough space for the meetings we're anticipating with more than a dozen people representing other food-community efforts in the area. I shouldn't have worried.

Since I last visited Madison, Willy Street had expanded, big-time, moving into spacious new facilities with a sunny conference room. This co-op, along with others in Madison, is one of more than 300 food cooperatives that have been sprouting up across the country since the '70s. (And this in the face of a tightening retail control that may break all records: In the three final years of the last century, the five biggest grocery retailers doubled their share of the market to 42 percent of all sales in the country.)

On this, our last day, we want to learn about the farming business, and specifically marketing. All the farmers we've met are inspiring, yes, but are they economically viable? While spending on food in this country since I wrote Diet for a Small Planet *has jumped from roughly $100 billion to $500 billion a year, the amount going to farmers has only crept up, with virtually no rise since 1980. Today only twenty cents of our food dollar goes to the farmer, down from forty-one cents in 1950—the rest goes to all the other stuff, from advertisers to packagers to distributors.*

Earlier, family-farm defender John Kinsman had said that part of the answer was to eliminate the middleman. Great idea, I thought, but how?

Seeming to answer this exact question, into the co-op walks dairy farmer Jim Miller, fit-looking and fifty-something with graying hair and a square jaw. After a tight-as-a-vise handshake, Jim sits his husky frame down and begins to tell us about Organic Valley, a cooperative of family farms working together to market and distribute organic products. "It used to be called the Coulee Region Organic Produce Pool," he says, "CROPP, for short."

As soon as I hear "CROPP," my heart sinks. I flash back to a small wooden church in Viroqua, Wisconsin. Over a decade ago, I'd been invited to speak there by the nicest, most earnest people—CROPP founders. They had just started an organic dairy co-op to cut out the marketing middlemen, and they had big dreams. Mainly, though, I remember sitting on a hard pew thinking how unlikely it was that they would actually make it. Assuming the co-op today couldn't consist of more than a mélange of well-meaning idealists, I'm afraid that, no matter what Jim tells us, I'll conclude that farmers' marketing efforts are marginal in the big picture.

But Jim is hardly a hippie back-to-the-lander. He was born and raised on the farm, left for what he called the "restaurant business" in the south, but came back to the farm in 1994.

"My father had just died of cancer. He was the one who did all the spraying of the fields. It was a terrible death. We couldn't prove it, but everyone in the family blamed the farm chemicals.

"That's when we decided to go organic, all of us—

ten families. We're all related and we farm together. Neighbors told us, 'You're nuts,' and we did have to suffer through some trying times, because it takes three years without chemicals before you can sell as certified organic—and get a premium price.

"I remember the ridicule. As we were shifting to organic, our corn didn't look so good, and one neighbor said, 'Hey, Jim, what are you growing there—pineapples? I've never seen such lousy-looking corn.' Now, farmers who scorned us are asking how they can do it.

"This summer, five fields owned by neighbors of ours are completely dead from the leaf hopper. A neighbor sprayed two or three times to kill the bugs, but it didn't work. The insect damage went right up to the edge of our fields, but our crops are as healthy as can be. Healthy plants in healthy soil will not be killed by bugs. Bugs scavenge for unhealthy plants."

I shouldn't have worried about that tiny, rag-tag group of founders I met in the church that evening in 1989.

Organic Valley membership has jumped from seven farmers to 300, and last year they sold almost $80 million* in organic products from California to Maine—now even Japan. If it continues to grow as fast as it has, Jim tells us, Organic Valley could reach sales over $100 million next year**—a far cry from the humble beginning I witnessed.

"More farmers want to join us all the time. The market is growing like crazy." Jim says, smiling, and he's right. Nationwide, in each of the last eight years' sales of organic food have grown by 20 percent.

"Consumers are smarter now. Once a week there's something in the paper about what's going on—hormones, pesticides, resistance to antibiotics. So the big food chains are responding. They are working with us to get organics into their stores. They are giving deals to companies like Organic Valley because consumers have a vote and they're using it."

I think of what John Kinsman said several days earlier: "With every dollar you spend, you vote. You either vote for big business, or you vote for the family farm." And, I think now, you either vote for big chemical companies or you vote for healthy farmers.

Jim tells us how satisfying it is to learn more each year about better, more sustainable farming practices—instead of just taking instructions from corporate advertising. I flash to farmer suicides, from India to France to Wisconsin, and am stunned by the tragedy of such wasted life and the contrast with Jim's experience. As we're saying goodbye, Jim looks at Anna and me. "You know why we're so successful?" he asks with a broad smile. "Because we love what we do."

{ Founding member of the Hamburg-based World Future Council, Frances Moore Lappé is the author or co-author of 18 books, including *World Hunger: 12 Myths, EcoMind, Changing the Way We Think, To Create the World We Want* and *Diet for a Small Planet*, her book that sold three million copies. }

Lappe visits CROPP, 1989

* In 2012 CROPP membership had grown to 1814 farmers, and total sales neared $860 million.

** 2013 sales projected near $1 billion

3. A Fair, Stable Price

1988–1990

at the Farm Gate

* They won't let you * A working mode

* Writing a constitution * Co-Dependency *

The "big boss" of the National Farmers Organization in Wisconsin was Steve Pavich. His NFO chapter had produced more money for the national organization than any other state, and the buck stopped at Pavich's desk in Sauk City, about 75 miles east of Viroqua.

Pavich was powerful. With his no-nonsense, get-things-done style, visitors knew they better be over-prepared for a meeting with him.

George Siemon and Spark Burmaster had an appointment to meet with the big man in the summer of 1988. CROPP's NFO champions, Adolph Trussoni and Mark Michel, had made a strong case for the fledgling organic cooperative called CROPP, and they advocated NFO's sponsorship of the dairy pool. CROPP's mission underscored what the NFO had long believed: Family farmers deserve a fair, sustainable price for their production.

But no matter how noble the mission, CROPP's dairy pool would go nowhere without NFO's blessing and support. CROPP lacked credibility on its own.

Siemon and Burmaster ambled into Pavich's claustrophobic eight-by-eight foot office. Pavich was chain smoking, and the windows were closed.

"Would you be willing to put out your cigarette?" Siemon asked politely. "My friend Spark here is allergic."

There was a long pause. Anyone who knew Pavich would have scoffed, "Those two bumpkins came in here to get money from Steve Pavich and they're telling him not to smoke? Are they nuts?"

> "Steve Pavich was a World War II vet, a farmer, a guy with great tenacity. I think he saw that the founders of CROPP had fire in their bellies, too." Mark Michel, NFO field representative

← Carol Anne Kemen (facing camera) and her coworkers cut and wrap organic cheese in the Cheese Room at CROPP's first La Farge, Wisconsin, headquarters in 1990.

Healthy Co-Dependency

As it turned out, the outwardly-gruff Pavich supported CROPP's organic dairy pool plan. "He was a good guy," Siemon recalls. "He offered to create an NFO organic program in partnership with CROPP on the condition that we manage every aspect of it. The NFO's role was to write the milk checks for our farmers, fund those checks up to $50,000 each to begin with, and cover our milk processing and start-up costs. The NFO became our enabler." Basically, the NFO fronted the needed cash flow and CROPP's collateral was its cheese.

Pavich put Siemon, Wayne Peters, one of CROPP's first dairy pool members, and Greg Welsh, CROPP's first paid manager, on the NFO payroll.

> "The NFO was our financial angel. Nobody else came close." Jerome McGeorge, CROPP's first CFO

NFO became a champion of CROPP when they needed it most. "All of a sudden," Siemon recalls, "when I went looking for an organic cheesemaker, I was no longer some unknown guy with a wild idea. Now, I was some unknown guy with a wild idea *and* NFO backing. This partnership carried the day for CROPP. The NFO fronted the big costs and without them, we never would have survived."

It was a healthy co-dependency, one that CROPP clearly needed. After all, CROPP started with a dream, but no resources. "Organic was our focus," Siemon says, "but we had no knowledge of most everything else. We had to find a way to work with the existing agricultural infrastructure." The NFO's support started CROPP down a path of building a business that was dependent on outside services. They hired someone to haul their milk, and someone else to make their cheese. "That was the start of building our business based on a co-dependency on the resources of others," says Siemon. "In modern-speak, we were a virtual business." Co-dependency has its risks, to be sure, but CROPP's evolving story would prove the benefits were far greater.

> "The farmers are dependent on CROPP's employees, our partners, our customers and all the others who have helped our dream come true." George Siemon

There were times when NFO leaders wondered if Steve Pavich was right about this CROPP gamble—especially after Pavich's untimely death early into the NFO/CROPP program. "I remember a meeting in Viroqua when several NFO people attended," says Wayne Peters. "It was kind of our 'coming out' meeting. Adolph Trussoni sat at one end of the table. Jerome McGeorge was on one side with no shoes on and George Siemon was on the other side with no shoes. We were trying to explain to the NFO how this little organization with a dream was going to revolutionize American agriculture."

"If we ever get this thing going," Peters later grumbled to Siemon, "I'm gonna buy you a new pair of shoes!"

There was also the matter of a mysterious file at Pavich's office with CROPP's name on it. When the organic dairy pool was Steve Pavich's project, no one asked questions. But with his death, the NFO staff got to worrying about several hundred thousand dollars the NFO had already invested. Siemon and McGeorge traveled to NFO's headquarters for a come-to-Jesus meeting with their financial angels. "They told us they looked for the CROPP file in Steve's desk to review the agreement with us," says Siemon, "and they said, 'The file was empty.' " This was resolved by Wayne Peters and George Siemon signing a monster stack of papers, changing nothing in the relationship, but making the NFO lawyers happy.

As if that wasn't disturbing enough, McGeorge told the NFO that CROPP kept two sets of books but—trust us, it's all on the up and up. How many times has two sets of books had the stink of fraud? But McGeorge had a reasonable explanation.

"The NFO kept our organic cheese inventory on their books at the conventional milk market value, even though they paid the higher organic milk price," McGeorge said. "They couldn't grasp the idea that organic had more value," and so the NFO took the conservative route in their accounting. They used the conventional milk price. "That created a problem for us because we were losing money on paper," says McGeorge. "The NFO wouldn't value our organic cheese at what they actually paid for the milk. The second set of books that we kept for CROPP's overall operations showed our cheese selling at the higher, organic price.

← Steve Pavich, the "big boss" of Wisconsin's NFO; George Siemon (left), Harriet Behar and Jerome McGeorge (in green) meet with Jeff Ward at the Swiss Chalet in La Crosse, Wisconsin, 1989.

Nate Trussoni displays his youthful NFO support, 1989. →

Accounting rules state that whatever you pay for a product is the cost of goods. Simple as that.

"The NFO got very upset," says McGeorge, "and even more upset because we admitted it. We just needed to represent the financial situation accurately."

By the fall of 1988, George Siemon was back to milking dairy cows and his was dairy pool farm family #8 behind the original seven families: Engel, Green, Hass, Peters, Rathe, Slama and Wedeberg. The farmers met weekly as volunteer managers of the organic dairy business while Greg Welsh, a former Iowa farm boy, managed CROPP's produce pool, and Margit Kaltenekker, a young botanist, coordinated CROPP's field and harvesting operations. The CROPP board of directors and an advisory council—all farmers—kept a keen eye on things. In February, 1989, Keith Johnson, a former teacher, Vietnam War vet and sheep farmer, became CROPP's first general manager, making $80 a week for his "part time" job.

> "The Wisconsin Department of Agriculture said a stable pay price would never happen in the world of dairy. CROPP's pay price is both stable and sustainable."
> Jerome McGeorge

Got Steak?

Early on, some of the dairy pool members never believed they would get paid $17.50 per hundredweight (one hundred pounds) for their organic milk, as Siemon had predicted. "If I ever get paid $17.50," Wayne Peters

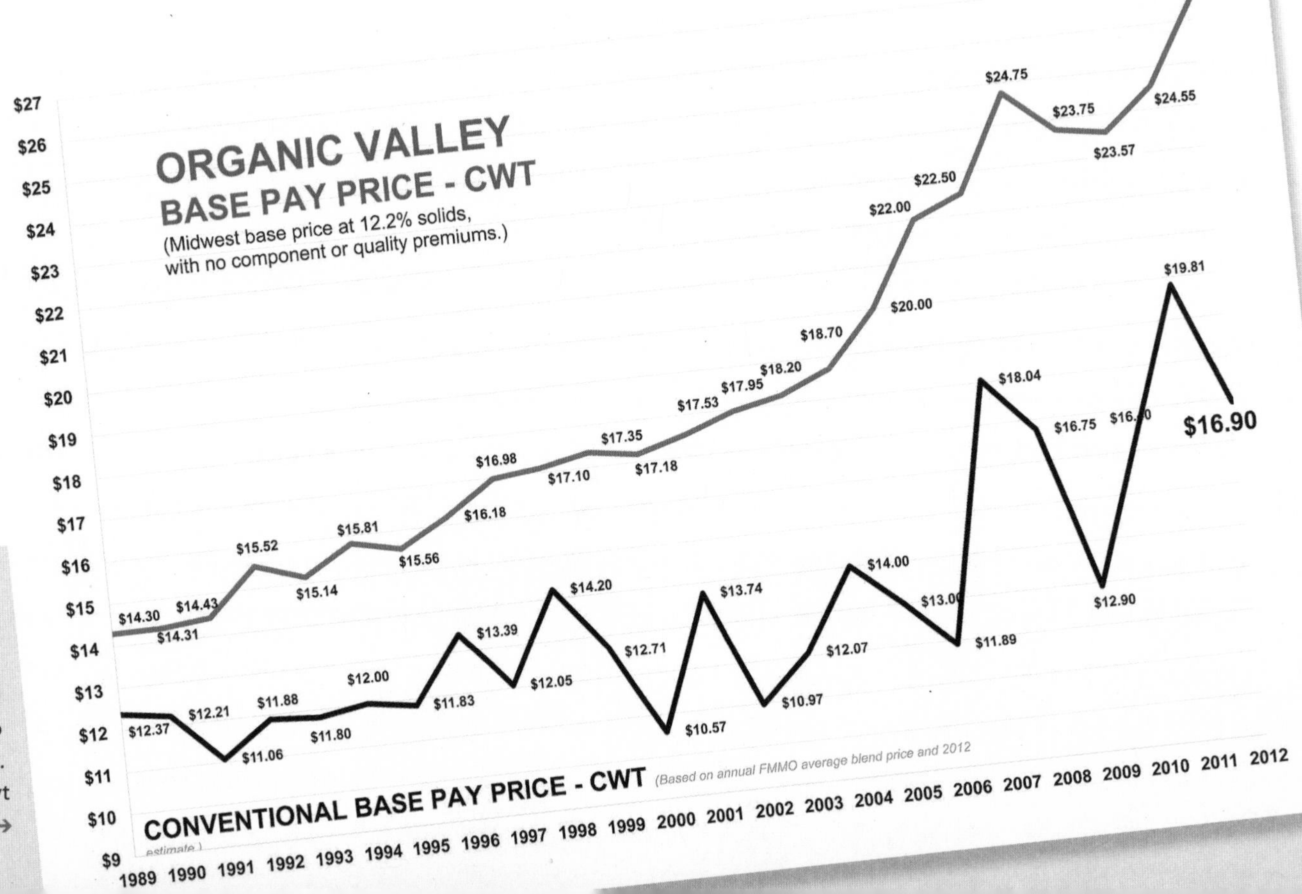

This graph from 1989 through 2012 shows the base price organic farmers earned for their milk, compared to conventional dairy farmers. The price is based on one cwt (100 pounds) of milk. →

told him, "I'll buy you a steak dinner!" (25 years later, Siemon kids Peters that he is still waiting for the dinner *and* the new shoes Peters promised. CROPP reached a Midwest base pay price of $17.53 in 1997).

Peters and Siemon became CROPP's "odd couple" who approached decisions often from very different perspectives, and yet, after vociferous debates, agreed most of the time. They were nicknamed "Ace" and "Slick," although it was never clear who was who. "We got pretty lively, and often Wayne and I were at the center of the liveliness. I can't figure out how he thinks to this day," says Siemon, "but he usually comes up with the same answer I do."

"The whole dairy project has been one of intense dedication by the Dairy Pool group with a strong faith and trust that it will succeed," Siemon wrote to CROPP members in August, 1988. "To me, the project has been a shining example of what people can accomplish when they work together and pool their resources."

Some members of CROPP's produce pool had mixed feelings about the dairy boys. Some were ambivalent and others didn't want them in the Cooperative at all. But no one could ignore CROPP's good fortune when organic dairy sales took off in the late 1980s and Wisconsin's awful drought stunted organic vegetable production. In years ahead, the financial success of CROPP's dairy pool would make other organic pools possible with early injections of start-up money. This proved the value of the cooperative spirit, as Jim Wedeberg, one of the original dairy pool members, said, "CROPP was conceived not for the benefit of any one individual or pool, but for the benefit of the whole."

> "We started CROPP with the concept of being a multi-product co-op. The only limit was this: Our products had to be organic and from the farm." George Siemon

Powerful Cooperation

In its first-ever annual report covering 1988, carefully hand-lettered by CFO Jerome McGeorge, the review of CROPP's accomplishments amazed many. More than 60 organic vegetable growers and dairy farmers had joined CROPP. The Cooperative provided its members with organic fertilizers, seeds, plants, supplies, soil testing, management assistance and a series of classes focused on commercial organic production. CROPP arranged for storage, refrigeration and processing for its members' production, and it developed a modest transportation and distribution system with a rented truck that cost $420 a month. A group of CROPP farmers, led by Dave Engel, started a local chapter of OCIA (Organic Crop Improvement Association), an international certification group. CROPP members also drafted the first organic standards for Wisconsin

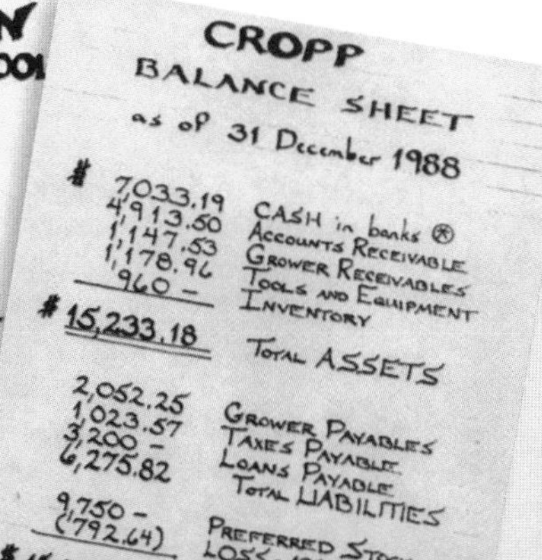

COULEE REGION
ORGANIC PRODUCE POOL
CROPP
1988
ANNUAL REPORT

CROPP
BALANCE SHEET
as of 31 December 1988

$ 7,033.19	CASH in banks ®
4,913.50	ACCOUNTS RECEIVABLE
1,147.53	GROWER RECEIVABLES
1,178.96	TOOLS AND EQUIPMENT
960 –	INVENTORY
$ 15,233.18	TOTAL ASSETS
2,052.25	GROWER PAYABLES
1,023.57	TAXES PAYABLE
3,200 –	LOANS PAYABLE
6,275.82	TOTAL LIABILITIES
9,750 –	PREFERRED STOCK
(792.64)	LOSS - 1988

← Keith Johnson, CROPP's general manager; Greg Welsh, CROPP's first paid manager, at the Hill & Valley Warehouse near Viroqua, a distribution point for CROPP's organic produce

Try It, You'll Like It!

Harriet Behar, the Wisconsin weaver and organic gardener, was a natural foods pioneer. She sold her fruits and vegetables the first year the Dane County Farmers Market in Madison, Wisconsin, opened in 1973, and she supplied local stores offering natural foods. She hawked zucchini to help out a friend in downtown Madison, and her relentless sales pitch earned her "the zucchini lady" moniker.

When she heard about the founding of CROPP Cooperative, this petite life force—more energized than the iconic bunny—phoned Greg Welsh.

"I think what you're doing is absolutely great," Behar gushed. "I'll do anything to help. I'll lick envelopes. I want to see this go!"

"Check back this fall," Welsh, the overburdened CROPP vegetable coordinator, answered. "You don't know who you're turning away," Behar sighed. But come fall, she was back and Welsh soon found ways to put her energy to work.

At first, in the fall of 1988, she worked with CROPP's vegetable farmers, advising them on organic methods, crop choices and handling, insect issues and estimating yields. She served as CROPP's own "ag extension agent" and her pay, like everyone else's, was five dollars an hour. When Welsh went out of town for two weeks, he left Behar with a Rolodex and the telephone. "Sell the vegetables," he said. "You'll do just fine."

"I felt like a used car salesman. 'What do I have to do to get you to buy two pallets of organic sweet corn?'" Harriet Behar

On her first day, Behar sold nothing. "We had a cooler full of vegetables and I was freaking out," she recalls. "Green beans, winter squash, zucchini, onions, corn, parsnips, tomatoes, bell peppers, lots of broccoli. I went home to Gays Mills at the end of that long day and I cried myself to sleep." The next day, Behar was determined. "I woke up and said, 'Damn it, nobody is going to say no to me today!'"

"I went around and asked our farmers, 'What makes your milk different? If I'm going to sell people on this, I've got to tell them why it's different.'" Harriet Behar

Behar proved to be so effective at reading the organic market and selling CROPP's vegetables that, in 1990 the Cooperative's eight dairy pool members each chipped in $100 to finance her month-long, barebones trip to California, the mecca of organic. She called on the owners of natural foods stores. It became clear to Behar that CROPP's organic cheese could be a big seller, but it needed a catchy,

CROPP REPORT # ELEVEN - AUGUST 18, 1989

VEGETABLE REPORT

by Harriet Behar

The vegetable rush is upon us and I'm sure that we all appreciate the abundance of fresh vegetables and fruits now before us.

When you think you are between 1-2 weeks from delivering produce please call me, Harriet at 872-2487, between 7-8 a.m. or 9-11 p.m. to catch me near the phone. You will be responsible for contacting me and either picking up your first load of boxes at La Farge or arranging for a drop off if the truck happens to be in your neighborhood. After the first load of boxes whenever the truck comes for pickup of your produce we can bring the boxes you will need for your next delivery as long as we are notified of what you will need.

The refrigerated truck will be providing pickups free of charge for loads of over 20 boxes on Tuesday and Friday. These boxes must be ready for shipment or there will be a fee charged for packing at the warehouse. The growers are also strongly encouraged to deliver produce and pick up boxes at the La

↑ Harriet Behar works the phones in 1994.

Harriet is a natural born saleswoman at the Natural Foods Expo in 1990; Dan Meyer and Harriet Behar, with Jane and Clovis Siemon, are the faces of CROPP at the 1992 Expo. →

memorable brand name, not a long name that was hard to remember.

"If I could sell hand-woven wool sweaters to people in the summer," she says, "I figured I could sell cheese." By the end of her first visit, Behar had met with major natural foods distributors, including Mountain Peoples (today part of United Naturals), Rock Island, Nature's Best, Northwest Naturals, Nutrasource and Rainbow Foods, as well as more than 50 natural foods retailers.

One year later, when Behar returned to California for the Natural Foods Expo in Anaheim, CROPP couldn't afford a booth. Cooperative folklore has it that Behar wore a many-pocketed trench coat and stalked the expansive exhibit floor offering tasty cheese samples to everyone. In fact, she carried cheese samples in her handbag ("Cute, cheap, but not very effective," she says.)

By 1992, CROPP had its own booth at the Expo, and Behar was in sales heaven. She had a great little sales sidekick: young Robert "Clovis" Siemon, age 12. The naturally gregarious child wore a cheese tie and a sign on his back that read "Follow me to booth 570a." He carried a full platter of Organic Valley cheese samples and, like the pied piper, he lured people to CROPP's booth and the indefatigable Harriet Behar.

← Before CROPP established its own Organic Valley brand, Harriet Behar successfully marketed the Cooperative's Coulee Region Organic Cheese.

in cooperation with national and international organic organizations, the Wisconsin state legislature and local universities. CROPP's board decided to require all farmer-members to be certified organic in 1989. (The U.S. would pass the Organic Foods Production Act in 1990, but it would be 12 more years before the national organic standards were fully implemented.)

That first year of CROPP's life underscored the power of partnerships—especially sponsorship of the organic dairy pool by the NFO and partnership with North Farm Cooperative of Madison, the organization that handled marketing and distribution of CROPP's organic cheese nationally.

And though the Cooperative's bottom line showed a $792.64 loss in the first year, it also posted $43,889 in vegetable sales, $51,086 in dairy sales (used to make five kinds of cheese), and $11,819 in revenues from the sale of seed, supplies and services to members. CROPP's organic dairy farmers earned $12.25 per hundredweight for their milk, about the same as conventional milk that first year. By 1989, however, the organic price would climb to $14.30, nearly $2 above conventional milk. The organic to conventional milk price difference would consistently widen over the decades.

If it weren't for the cooperative spirit among an army of volunteers, CROPP would have been deeper in debt. The Cooperative paid only $15,380 in labor and payroll taxes in 1988 and just $4.50 per hour to a few part-timers. Everyone else just pitched in because they believed in CROPP's purpose.

Got Chicken?

There was no line item for organic meat sold in 1988, though CROPP intended to market chicken grown with organic feed and no chemicals. "The meat had been processed in Iowa and labeled with an organic sticker," Jim Wedeberg says. "When the USDA examined the chickens in Madison, Wisconsin, they told us the stickers had to be removed from all 25,000 pounds of chicken." (That was before meat certification rules actually existed.)

OK, we'll remove the stickers, CROPP agreed. "But," the USDA commanded, "you'll deliver that chicken back to Iowa where the meat was processed, remove those stickers and *then* bring your chicken back to Madison for sale." Oops.

"We were," Siemon admits, "dangerously unrealistic in those days."

Unrealistic, perhaps, but there had to be room for dreams and aspirations in CROPP's early years. Marketing CROPP's organic cheese was a case in point.

When Torkelsen's Springdale Creamery in Readstown, Wisconsin, produced America's first organic cheese made with organic milk from CROPP's original dairy pool members in 1988, the young cooperative was suddenly in the cheese business. To monitor quality, CROPP chose to cut its own cheese in its old headquarters building. Board member Wayne Peters scouted around for used cutting equipment, convinced that CROPP had to be hands-on to make sure this new cheese line was high quality. Though CROPP had made it a practice to rely on outside partners to provide services and skills, the Cooperative still had a stubborn "we'll-do-it-ourselves" streak on certain key tasks, like quality assurance.

"Our milk is made into organic cheese to provide the consumer with a pure and wholesome product." The Hass family

U.S. organic industry sales reached $1 billion in 1990, a mere .002% of the total U.S. food industry.

Within a few months, North Farm Cooperative started marketing and distributing CROPP's cheese nationwide to a large network of natural food stores under its label, North Farm Organic Cheese. The line included five varieties made with raw or pasteurized milk: Low-Fat Cheddar, Colby, Brick, Muenster and Low-Fat Monterey Jack. CROPP tried to sell its own Coulee Regional Organic Cheese label to supermarkets with little success. Only some in nearby La Crosse and Viroqua picked it up.

"Organic cheese is made from certified organic milk produced by organic dairy farmers," the introductory brochure declared. "Organic dairy farmers feed their cattle grain and hay grown without pesticides and herbicides. They give no antibiotics or hormones to their cows." In a "Meet the Farmers" section of the brochure, CROPP's dairy pool families explained why they chose to be organic farmers: ". . . to regain the natural condition of the soil," Neal Green said. "In doing so, we have greatly improved human and animal health."

"This is a more productive way of farming for us," said Dean and Elsie Rathe. "It is easier and healthier for us, the animals, and healthier for the consumer."

"Producing organic cheese," the Peters family said, "is one way we can ensure our family and the nation a healthier environment."

George Siemon called the cheese line "a brave effort by a handful of farmers" with the support of CROPP, the NFO and North Farm Cooperative. "They have been frustrated with selling their high quality organic milk through conventional markets where any possible appreciation of their organic methods was lost," he said. "They are tired of a pay price too close to bankruptcy with the additional irritation of being dependent on government support. They believe the consumer wants the choice of organic cheese."

Playing Out of Our League

At first, CROPP's cheese varieties tasted a bit too similar. "We used to joke that we had one flavor—'white'—and you could call it cheddar, you could call it colby, you could call it aged cheddar," says Jim Pierce, an early CROPP part-time employee and today Global Program Manager for Oregon Tilth, a leading organic certification agency. While CROPP's organic cheese would, in time, win awards, Pierce says, "It all kind of tasted the same at first."

CROPP's cheese had to jump through a few hoops before it could go national. For one thing, it had to be "graded" by a cheese expert. Wayne Peters knew an award-winning cheesemaker named Florian Franke near Dodgeville, Wisconsin, who might be able to help.

Peters arrived at Franke's cheese shop with one block of every kind of cheese in the CROPP line. "Cheese has to be tested, so I asked him to do it," Peters recalls. "Then I told him about our plans to market our organic cheese nationally. I said we planned to do the same with organic milk."

> "Our cheese passed Florian's inspection. He never said it was bad. We thought it was good." Wayne Peters

Mr. Franke, the "cheese wiz," was incredulous. How could CROPP compete on quality against established cheese producers? He sampled a few of the blocks and waited for Peters to finish his summary of CROPP's ambitious plans.

"No, Wayne, tell me what you're *really* doing," Franke said.

"What we're *really* doing is selling the first organic cheese nationally," Peters repeated himself.

Silence. Franke tasted another sliver of cheddar and wondered, "Should I be blunt?"

"If you do this," Franke told Peters, "aren't you guys playing out of your league?"

Peters tackled the question head on. "You know, Florian, if we don't ever play out of our league, we're never going to get out of our league."

They Won't Let You

The founders of CROPP had heard plenty from nay-sayers. A common warning was, "They won't let you." CROPP's response was, "Who are *They?*" Practically speaking, *they* were the big boys of agribusiness—the Krafts and the ConAgras and the big meat processors like Hormel. "They seemed to have control of everything in the grocery store," Wayne Peters said. "People told us, 'They'll never let you do this.' I think we snuck in the back door and *they* didn't even know we were doing it."

"We heard, 'They won't let you' a lot," says George Siemon. "We trusted our vision and backed that faith with hard work. Maybe those who don't think something can be done should just get out of the way. I believed the way to change the world was to build a working model; not to go out and tell the world, 'You've got to change.'"

A version of "They won't let you" also came from North Farm Cooperative, CROPP's organic cheese marketer and distributor. Michael Schachter of North Farm thought he was simply acknowledging the hard realities of the marketplace when he suggested that CROPP farmers might be smart to reconsider their pay price. "You guys could sell a lot more cheese if you'd lower it," he told them. To that, Ray Hass huffed, "I'll pour my organic milk down a rat hole first!" CROPP stood firm on its price.

In that first year, North Farm sold about $44,000 of CROPP's organic cheese, and in just the first two months of 1989 alone, total sales spiked to $100,000. By October, 1990, CROPP was selling 30,000 pounds of organic cheese every month (about $80,000 monthly went to the NFO/CROPP partnership). One year later, CROPP had 17 cheese varieties sold by 12 new wholesale distributors all over the U.S. including Organic Farms, a company serving the Eastern U.S. These customers distributed CROPP's cheese to manufacturers and retailers in 20 states and on both coasts. CROPP also sold its cheese to private label customers: Tree of Life (they were piloting organic pizza), Earth's Best (organic frozen dinners) and Walnut Acres (for their popular organic mail order business). North Farm—once the sole marketer and distributor of CROPP's cheese nationally—had become the Cooperative's biggest private label customer, instead.

Selling the Story Behind the Label

North Farm had many clients, and CROPP's line was only one line among many. CROPP believed they could do better on their own and ended North Farm's exclusive marketing and distribution relationship. To tap into more markets, the Cooperative recruited other wholesale distributors and private label customers and—most important—decided to create their own brand.

Thanks to a $20,000 Agricultural Diversification Grant from the Wisconsin Department of Agriculture, CROPP had funds to create their brand, design the labeling and build a promotion and distribution program. Mark Kastel, a former farm equipment dealer who grew up in Chicago, but quickly turned to farming, signed on with CROPP to conduct market research and help create the brand identity and marketing program. Kastel made $5 an hour, just like everyone else. (There were times, he says, when CROPP's leaders didn't get paid because the ladies cutting cheese for CROPP needed the money more.)

In late 1989, a brainstorming team gathered inside CROPP to begin work on the new brand identity. Among the participants were Keith Johnson, CROPP's general manager; George Siemon, CROPP's dairy pool coordinator; Harriet Behar, vegetable coordinator; Jim Pierce, jack-of-all-trades and truck driver; Sally Marshall, CROPP's new office manager; and Kastel.

"The best name you can come up with is something that conveys what you're all about," Kastel told the group, "so put organic in your name. People aren't just buying cheese or your label, they're buying the story behind your label." The group fired off ideas and two of them were "Organic Valley," suggested by Behar, and "Family of Farms," offered by Pierce.

The group had reached a creative impasse in their work, Pierce says, when he excused himself to drive to La Crosse for a load of cheese. As he traveled, he

cheesemaking plant

Jeff Ward, a third generation cheesemaker, holds a finished round cheese block.

Christine pushes the cheese hydraulically through a series of wire harps which can be adjusted for various weights and sizes

A smiling Carolann bags cheese (and weighs if for exact weight cut pieces)

↑ Photos from CROPP's 1989 "cheese education" photo album show it was serious business.

considered the names suggested. When he reached the La Crosse cheese plant, he phoned CROPP and got Marshall on the line. "Here's my idea," Pierce said. "Draw an oval, write Organic Valley along the top, write Family of Farms along the bottom and put a pastoral scene in the middle."

Bingo. Pierce's brainstorm became CROPP's first brand mark and the label design continued, with few variations, for years. When Pierce told Marshall, a newcomer to CROPP, that she'd have to start answering the phone, "Organic Valley, this is Sally," he jokingly predicted that the little town of La Farge, Wisconsin, CROPP's new home, might one day be renamed Organic Valley. "Who knows," he told Marshall, "We're going to be huge, you know."

First things first. CROPP applied for a trademark for Organic Valley Family of Farms and chose Mixed Media of Madison to design the label and Belmark Label of De Pere, Wisconsin, to print them. The brand first appeared in April, 1990, on Organic Valley cow and goat cheese, frozen organic corn and vegetable boxes.

In 1992, a National Cancer Institute study found farmers have higher rates of certain cancers, including leukemia and prostate.

A Working Model

Some people who thought CROPP's revolutionary model of agriculture *was* viable became shareholders when the Cooperative had its first preferred stock offering.

"We have a broad base of support in our growers, from high school students to 70-year-old farmers. We have Amish, full- and part-time farmers, tobacco growers and experienced vegetable producers," the stock offering letter said in 1988 and 1989. CROPP laid out its mission, goals, definition of organic and the foundational principles that would guide the organization. They wrote their own "constitution" that defined the DNA of CROPP and they shared it with future stakeholders.

"CROPP is a true grassroots movement in response to the needs of both the farmer and the consumer in the quickly developing organic market," the letter said. "We're not promising a pot of gold, but we are determined to do something about the seemingly hopeless problem of the deteriorating farm culture and economy."

They were honest. "With the present depressed farm prices, it's very hard for us to be self-financing." For one thing, grants and government loans that CROPP had applied for were delayed in bureaucratic paperwork. "We have elected to sell preferred stock to raise funds so that we are able to purchase equipment, pay salaries and pay overhead costs," the offering notice continued. "We are asking CROPP growers, neighbors, businesses, churches, other cooperatives and all concerned people to buy preferred stock to help us as a sign of community support. It is CROPP's intent to buy back these preferred stocks in future years. We feel CROPP has great potential and will be a welcomed boost to our economy in years to come."

> "Please help us now in our infancy." CROPP letter to potential investors

CROPP's farmer-members each pledged to sell $200 worth of preferred stock at $50 per share. Each year, the stock offered an annual dividend of up to eight percent of its par value.

To sell the stock, CROPP needed stock certificates and the task fell to Spark Burmaster. "I got out my computer, typed out the words and went to a copy store," he recalls. "'You got borders?' I asked them, so we got borders. 'You got colored paper with texture?' We got nice green paper, ran it through a copy machine and cha-ching, we had preferred stock certificates. There were typos on a few of the early ones. Of course, the preferred stock sales staff, that's us, had the real job of selling it. Twenty-five years later in 2013, 100 percent of the preferred stock holds its stated value."

It Takes a Village

The community had already rallied behind CROPP, so a successful stock sale looked promising. A group of musicians and storytellers staged the Kickapoo Old Time Radio Show Fundraiser and raised about $400 for CROPP. When they turned the money over, they simply said, "Use it the best way you can." Local churches chipped in cash. The Wisconsin Development Corporation gave CROPP a modest seed grant. The Wisconsin Technical School designed organic vegetable production classes for potential CROPP growers, even though the school never had the course in their curriculum before. Dahl Pharmacy in downtown Viroqua offered free office space.

The Mayor of Viroqua, Chuck Dahl, was among the first to commit $1,000 for 20 shares of preferred stock. Margaret Siemon was the largest, single investor when she anteed up $5,000 for the cause. Her son was taking her advice to "serve the world"—it was the least a mother could do.

The investment turned out to be a good long-term bet for those original investors, though the young CROPP faced the usual cash flow and financing issues that torment most start-up ventures. By 1989, CROPP's gross sales rose to $412,022 (from $106,794 in 1988) and then took another leap to $723,696 in 1990. Cheese was king. It represented 84 percent of CROPP's gross sales. The vegetable pool struggled, though its total sales had increased at a respectable rate. Even so, the program still lost money. While CROPP's dairy program had NFO support, a marketing outlet in North Farm and plenty of local cheesemakers to hire, the vegetable pool had none of these benefits. They had no financial angel, no national distribution options and no conventional market to turn to if organic purchases fell flat. The vegetable pool members faced these big issues on their own and created a business from scratch with little experience. It was a tough row to hoe.

CROPP's organic meat sales were below expectations, too, and CROPP's board did some hard soul-searching. Should they drop out of organic vegetables and meats? They chose patience.

You've Got a Friend

The Cooperative posted modest losses for 1989 and 1990 as the costs of growth weighed heavily on CROPP's income sheet. They made plans to seek new financing and found a friend in Bill Bosshard at the Bank of La Farge.

"CROPP wanted to buy the Warner Creek Cheese Factory in La Farge that had been closed and sold to AMPI (Associated Milk Producers Inc.)," says Bosshard, a second-generation banker whose father, John, an attorney and farmer, had interests in several small banks around Wisconsin. "CROPP was going to be active in our town and that purchase started our relationship." Bosshard knew Wayne Peters: "He was a solid farmer on the ridge in Coon Valley. He was as good as there was. He wasn't a dreamer from the city. Wayne believed CROPP was the salvation for the small farmer."

Peters introduced Bosshard to George Siemon, Jim Wedeberg and Jerome McGeorge. It helped that Peters and Wedeberg always wore shoes and looked like regular farmers. The fact that Siemon had long hair and McGeorge often took cues from astrological signs to suggest the best times to meet with bankers didn't seem to bother Bosshard. "These guys were smart enough and very dedicated," he says. "They wanted to work seven days a week. To me, the project was worth taking a chance on."

Bosshard remembers climbing "those creaky stairs" to the second floor of the empty cheese factory in La Farge in 1989. "It was a building that wasn't really of much collateral value," he says. "It was in the flood plain and it was old and outdated, but CROPP made it work for many, many years. They did a lot of the upgrading themselves. From the start, CROPP was smart because they didn't tie up a lot of money in their physical plant."

Creative Financing

Keith Johnson, CROPP's general manager, had found the La Farge building on the west end of the sleepy, little village. It had been on the market for several years. The price was well below any property CROPP might buy in nearby, progressive Viroqua.

AMPI had paid about $250,000 for the building originally, but they offered it to CROPP at a fire sale price: $24,000. AMPI just wanted it off their books. CROPP didn't have the 10 percent down payment, but the factory *did* have a 5-foot by 30-foot stainless steel cheese vat that went with the property.

Dairy pool member Wayne Peters arranged to sell that vat for the same amount as the down payment. On the day that CROPP's officers signed the contract for deed to buy the plant and they turned over a (momentarily) bad check for $2,400 to AMPI, an equipment broker handed Keith Johnson $2,400 for the cheese vat. He raced to the Bank of La Farge to deposit it. (As it turned out, AMPI had to pay to remove three fuel tanks from underneath the building, so their net return on that sale was zero).

In classic can-do style, Wayne Peters rounded up other CROPP farmers to help haul that bulky, two-ton cheese vat out of the creamery. "Wayne loves to solve problems," says Spark Burmaster, "especially if it involves concrete and steel."

> "We were building CROPP from nothing."
> Keith Johnson, first general manager

CROPP's first, real home (every place before that was borrowed or leased) was a mix of sturdy stone masonry and wood construction with a cheese room, two walk-in coolers, a loading dock, garage, office area and storefront with a counter and display coolers. There was an apartment upstairs, once used by the cheesemaker and his family.

"We finally had a center, a place to work and collect organic vegetables and soon we were cutting our cheese in La Farge," says Johnson. "The existing coolers were almost functional. Wayne Peters and his sons came over

← Bill Bosshard, of the Bank of La Farge, helped CROPP navigate its touch-and-go early years.

Keith Johnson found CROPP's first permanent home: a run-down cheese factory in need of TLC. →

and poured cement so we could move our cheese and vegetables back and forth and load them on our truck. I remember Jim Pierce up in the crawl space blowing in insulation. We converted the top of the building into offices. We got started on a shoestring and we were all pretty much working for nothing." Larry Jansen, a local refrigeration expert, knew that CROPP needed large scale industrial cooling units and he volunteered his time to see that the Cooperative was equipped. "That's a big reason why CROPP succeeded," says Spark Burmaster. "A lot of people believed in our mission and they stepped in to help."

> "The La Farge creamery was CROPP's home base and their total investment for many, many years. They were shrewd."
> Bill Bosshard, banker

In 1960, 17 million cows produced 14 billion gallons of milk. In 1995, 9.4 million cows produced 18 billion gallons.

CROPP's Board President Jack Pfitsch remembers using that cool cheese-cutting room for more than prepping the product. In the early years, they held annual meetings in that space with attendees sitting on folding chairs in a circle for hours. It wasn't long before the concrete floor and 55-degree air temperature had participants shivering.

The La Farge purchase turned out to be more than a building with potential. If home is where the heart is, then Keith Johnson had led CROPP's nomadic pioneers right to their heartland. Once settled in La Farge, CROPP began to find its core community, its footing, its family and its home.

Saturdays at the "Woodshed"

As CROPP gained farmer-members, their production grew and ample cash flow was crucial. For example, a dairy pool member had to be paid for his organic milk within 20 days, regardless of how long it took to convert that milk into aged cheese and sell it. In 1988, CROPP asked its farmer-members to voluntarily pledge a cash amount to help with this early cash flow issue through the Westby-Coon Valley Bank. Together, CROPP's dairy farmers pledged about $100,000 and the bank gave CROPP access to a $50,000 cash flow loan. If CROPP ever defaulted on its loan, the farmers were on the hook to pay, but that never happened.

Later, in 1993, dairy pool farmers were asked to invest 5.5 percent of their annual sales (about 20 days of milk production) into certificates of deposit with their names on it and CROPP's name. That money served as collateral for the milk checks, and it was the

BUILDING PURCHASED

In a late-breaking news development as this newsletter was going to press, the closing papers on the La Farge facility were signed on Friday, August 11th. This culminates a long search for a home for CROPP and many months of negotiations between CROPP and AMPI. The building committee of Spark Burmaster, Dan Stewart, and Jim Wedeberg should be congratulated on the work that they have dedicated to this process. Special gratitude should be extended to our manager, Keith Johnson, for his patience and dogged persistence in overseeing and expediting the long negotiating process. Thanks and good work!

← Big news in the August 1989 *CROPP Report*

The La Farge cheese room doubled as an early venue for potluck dinners among CROPP's farmer-members. →

Keep Calm and Carry On

Every business needs computers. In 1988, CROPP's entire computer system was Spark Burmaster's Commodore SX 64 (kilobytes of RAM). When Keith Johnson became manager in 1989, he brought his own souped-up Commodore 64 with two 5½ inch floppy drives. No hard drives existed then.

Before the Cooperative received a grant to buy its own Macintosh computer and printer, Dan Meyer wrote invoices by hand. Penny-pinching was rampant everywhere. Dan Hazlett remembers searching for used cardboard because they couldn't afford mats for the cheese cutters who stood on a cold concrete floor. The pair picked up organic eggs from CROPP's Amish farmer-members, transported them to La Farge for refrigeration and returned at 3 a.m. to haul the eggs 50 miles to Reedsburg Egg for packaging. The company wouldn't handle CROPP's tiny load after 5 a.m. when things really got busy.

They transported Amish farmer-members in an old red van to evening meetings at CROPP because the Amish don't drive cars. They cleaned, tabulated inventory, loaded pallets for transport and delivered CROPP's organic products to Milwaukee and Madison retailers.

(This page, left to right): Keith Johnson with his Commodore 64 computer in 1990; Mickey Keeley and the old red transport van; bike brigade: (right) Dan Hazlett and Louise Hemstead (CROPP's future COO) and Jim Kojola and Dawn Kelbel

On one run, Hazlett was rushing. "I wanted to reach all the retailers before they closed and my last stop was Magic Mill on South Park Street in Madison. I took a shortcut and drove underneath a railroad trestle. All of a sudden, the truck jumped a little and I looked in the rear view mirror. The refrigerated unit was on the road behind me. The bolts were so rusted that the box just popped off when it hit the iron belly of the trestle.

"We needed a flatbed truck anyway," Hazlett says, "and we used it for new construction after that!" They continued delivering organic products by strapping the loads down to the "modified flatbed." "We couldn't afford another truck," says Meyer. "The whole operation was seat-of-the-pants."

When CROPP held its grand opening at La Farge in 1990, the event was unforgettable. "I was a bachelor then," says Meyer, and the food CROPP served was delicious: roasted corn, pork chops, mashed potatoes." The original farmer-members were greeters, and Harriet Behar, CROPP's vegetable coordinator, led a tour. "She walked us into one section of the building and said, 'over here will be our milk trucks.' All we saw were cobwebs and a huge hole in the floor and I'm thinking, 'Really?'

"Here was this little company barely in existence and they had a grand opening. They were enthused!" Dan Hazlett

"Then Harriet took us through a hallway and said, 'This will be our cheese room.' We walked over pieces of concrete because Wayne Peters had been in there with his jackhammer. She led us into this smaller room and she pushed some electrical wires aside. Suddenly there were sparks and the lights went off in the whole building. Harriet never lost a beat the whole time."

Dedication of the La Farge headquarters (from left): Glen Alderman, La Farge Village Board president; Keith Johnson, CROPP's general manager; George Wilbur, president of the Kickapoo Valley Association; Senator Brian Rude; Representative DuWayne Johnsrud; Jack Pfitsch, president of CROPP's Board; Neil Jorgensen, University of Wisconsin College of Agriculture; and Will Hughes of the Wisconsin Department of Agriculture, with his son; (above, right) Dan Meyer in CROPP's cheese room

Hazlett, Meyer and Mickey Keeley remember the little break room at La Farge where folks could drink beer on their lunch hour (now a distant memory). Keeley joined a small group of women who cut Organic Valley cheese in the frigid cheese room. They were nicknamed "the pack of wolves."

"Those women cracked the whip," says Keeley. "They were intolerant of men who were slow and couldn't keep up with them. I never did feel quite good enough." Keeley says he advanced to "first official full time warehouse guy," then "first official mailroom guy."

CROPP was starting to outgrow its building and began renting workspace up and down the main street of La Farge. Keeley traveled to La Crosse and bought about 20 used bicycles. "We used those bikes to get around, deliver mail and visit our co-workers on our lunch hour," Keeley recalls. "We felt like the whole town was Organic Valley."

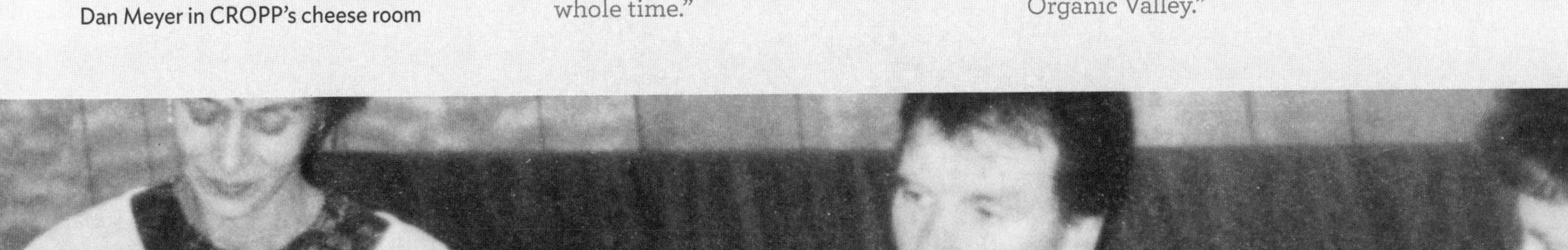

start of CROPP's long-standing member equity program. Bill Bosshard's Bank of La Farge took on farmers who needed financing to fund their CDs. "Over the years, CROPP was able to add farmer-members without borrowing a lot more from the bank because the farmers themselves either had cash to make the investment or they'd borrow it from our bank and the co-op would guarantee it," says Bosshard. "Self-funding has been a key to CROPP's success. They managed their cash flow and didn't get highly leveraged the way some co-ops do. They had the adequate working capital to accomplish the things they wanted to do."

These funds invested by farmer-members were converted to Class B stock in 1997 with a sweet annual dividend of eight percent.

CROPP was a reliable bank customer, but there were times when Bill Bosshard "took George to the woodshed," says Wayne Peters. This occurred at the frequent Saturday morning meetings with Bosshard at his bank office. "We didn't ever want to cut our pay price," Siemon recalls, "but Bill wanted to know if the going got tough, would we do it? We had this ongoing tension. Bill is the most common sense, practical guy about financing. He educated us."

"They had the right combination of social goals and dedication," says Bosshard. "But you still have to make a buck. All the good that they could get done would be for naught if CROPP didn't make it financially."

Bosshard credits CROPP's leaders with always emphasizing high quality organic production. "They pounded this idea into farmers' heads: if our products aren't high quality, they won't sell and you won't get paid. Farmers are tremendous producers, but they've never been the best marketers, and that's what CROPP provided—a tremendous marketing vehicle and a niche called premium organic. CROPP never had a recall or tainted product that hurt their name."

Three years later in 2000, after CROPP's financial needs had vastly outgrown Bosshard's Bank of La Farge and Bank of Bangor, the Cooperative hit a financial reef and briefly floundered. Bosshard would step in again and be the supportive, practical friend that CROPP needed.

A CROPP board meeting in 1990 (no worries about a dress code) →

First annual Wisconsin Organic Growers Conference in 1990 (from left): Keith Johnson, Dave Engel, Richard deWilde and baby, Ari, Margit Kaltenekker, Faye Jones, Eliot Coleman and Harriet Behar →

The Organic Persuasion * *Dr. Paul Dettloff*

"What is right will prevail." —*Voltaire*

After being a classical chemical veterinarian for more than half of my career, something surprising happened: my dairy clients started to go organic on me. I believe, to be a good large animal veterinarian, you have to be a good observer. So I sat back and watched.

These organic dairy farmers backed off on production pressure by feeding less grain. They fed cow's milk to the calves instead of milk replacer. They quit using hormones on their cows. They also were forced into a new fertilizer program for their soils, since organic forbade the use of synthesized fertilizers. They went back to short crop rotation and cultivating. They went back to using good farming practices that had stood the test of time.

What did I observe? A 75–80% reduction in veterinary work on those farms over three years, and it happened on nearly all of them. What a turnaround in the total animal health of the farm! It changed how I practiced medicine.

I also saw a philosophical change within each farm family. They became part of the ecosystem. They saw the deer come back to eat their non-GMO corn. The bird populations increased. The soil improved. It was at first subtle, but over time, quite dramatic. I often heard, "Farming is fun again, Doc."

* * *

Something else happened just before I retired from 35 years of active practice. I had a young farmer who was an insufferable Doubting Thomas, who had gone organic only because three of his close neighbors and farmer friends were organic. At first, everything had to be proven twice before he believed it.

He called early one morning with a cow down with milk fever. The dry cow lot was behind the barn a ways and we usually took his John Deere Gator out to treat cows in that paddock. As I came out of his milk house with my pail of warm water and sanitizer, he asked "How's your hip this morning, Dr. Paul?" (I eventually left practice because I'd worn out my right hip.)

I said, "Okay," (too proud to admit to the pain) as I limped to his Gator.

He asked, "Do you think you could walk that far if I carried your things?"

I said, "Sure I can. Why, is your Gator broke down?"

"No," he said, "but we had a good rain last night, and the drive is covered with night crawlers and earthworms. We'll kill a lot of them with the Gator, so I thought we'd walk."

He grabbed my pail and grip and took off for the cow. I limped along behind him, watching my step.

{ Paul Dettloff, D.V.M., has served CROPP farmers for more than ten years. A pioneer in organic large-animal veterinary practices and author of *Alternative Treatments for Ruminant Animals*, "Dr. Paul" has developed treatment protocols for many natural remedies, botanicals and homeopathic medicines. }

The Soul of Organic * *Harriet Behar*

At the heart of our choice to farm or eat organic, is the rejection of the many environmentally damaging methods and toxic materials used in conventional agriculture. Instead, we are amazed by both the simplicity and complexity we see in the natural world and choose to be led by the inspiration and durability of natural systems. It just makes sense to organic farmers that we should understand the workings of the natural world, and use what we have learned to then grow and produce food and fiber in a way that is regenerative and resilient, organic.

Many producers have been drawn to organic production due to its inherent demand for independence and self-reliance. Organic farmers rely on their own hard-earned knowledge and management to improve soil fertility, control weeds and humanely steward animals. They shun the chemical inputs and genetic modification pushed by multi-national corporations, and keep off the terrible treadmill that forces farmers to buy the next best thing *which promises to squeeze out a few more bushels per acre. Diverse crop rotations and/or grass-based operations, a respect for native plant communities and a poison-free environment provide food and habitat for diverse and plentiful wildlife, offering each organic farmer another reward: a front row seat at their own "garden of Eden."*

In my many years as an organic inspector and educator, over and over again I have heard the pride of accomplishment in the voices of farmers who work in harmony with the land they have the privilege to steward. In my own farming experience it is difficult to describe the deep sense of satisfaction when I see the positive effect my careful effort has had on soils, crops and livestock. A heavy early spring rain can be the cause of concern, but I have watched the water running off my field be free of silt, noting the cover crops' roots and biomass holding the soil together and abundant earthworm tunnels providing pathways for most of the water to infiltrate rather than cause erosion. Organic farmers hold many of the answers to our pressing environmental problems, including restoring balance to our climate and protecting our precious water and soil. Yes, organic agriculture can feed the world—and may be the only way to save it, too.

Organic farming is a cooperative endeavor, practiced by farmers who share a deep belief that we cannot successfully dominate our environment. A wise organic farmer once told me, "If we feel we are in a war with weeds and insects, then we have already lost." Instead, we seek out solutions and tools that are based in the inherent balance of the natural world's healthy ecosystems. Rather than concentrating dairy cattle indoors, providing rations that do not respect their rumen-based anatomy, concentrating their manure until it changes from a rich source of fertility to a pollution problem, we have found a better way. The elegance and beauty is evident when both land and animals benefit by organic cows feasting upon deep, green, well-managed pastures. We are learning to honor all living things from the smallest soil microbe to the living plants and the animals that feed upon them. At its core, organic farmers understand that we gain nothing when we use toxins to grow our food, but we lose so much from their negative impacts on the natural resources of water, soil and wildlife, not to mention cancers and other sickness in humans—disproportionally farmers.

Organic recognizes that everything in our world is interdependent. Everything.

CROPP Cooperative's first marketing slogan, in the early '90s, was "Farmers and Consumers Working Together for a Healthy Earth." It is this recognition that unites us: that our interactions with each other and our environment are the foundation of all change, whether for good or for ill. Many of us seek peace and meaning in our lives, with organic agriculture providing a sense of community for many farmers and consumers.

Hopefully, organic agriculture can continue to provide this haven outside the cut-throat world of conventional agriculture where, according to its own threatening maxim, you must "get big or get out." At farmer-gatherings I have often heard recently-converted organic farmers say how unusual it is to be with farmers who are happy, relaxed and engaged! Another fundamental tenet of organic agriculture, with CROPP leading the way, is that a fair price is due to the producer of the raw commodity. The economic welfare of the farmer is just as important as protecting the earth. By pushing for a fair return for their products, CROPP farmers set the foundation for a strong, healthy organic dairy "industry" in the United States. Organic consumers have always understood that they are voting with their dollars for family-scale farms.

At its essence, organic agriculture is a long-term view of food production and environmental stewardship.

While some farmers are initially attracted to organic production for the higher price they can receive, they stay with organic because they are "converted" to preserving and enhancing the health and productivity of their natural resources. As an organic inspector, I have visited farms their first year of certification. When I returned five years later, the farmer excitedly said, "I want you to see the incredible improvement in my soils, and how many earthworms I now have!"

The lifestyle of the organic farmer is not easy, but it is extremely rewarding for the soul, and, often, the only viable way a small farmer can be economically successful. The dominant agricultural world's love affair with "better living through chemistry" has made us all unwilling guinea pigs in a grand experiment, with organic the only option for those who would rather rely on what we know is productive, renewable and safe through millennia of experience.

Organic can provide food and fiber for the world's population by being attentive and respectful of each unique ecosystem. Organic farmers and advocates alike know it is the only way to sustain and nourish ourselves while protecting and enhancing our earth for the many generations to come.

{ Since 1989, Harriet Behar and her husband have operated a 216-acre certified organic vegetable and small grain farm in the hills of southwestern Wisconsin. Following her time as CROPP's first marketing director, Behar worked as an organic inspector for more than 15 years, and now as organic specialist and educator with MOSES (Midwest Organic and Sustainable Education Service). }

↑ Aaron Brin, Harriet Behar and Cookie, on their organic farm

CROPP Timescape

1983 -201

FARMY

1983–1987

National Context: 1985: FDA approves bovine growth hormone.

CROPP Cooperative:

- **1983–1986:** Annual (1–4) Driftless Bioregional Network gatherings *(pages 14–15)*
- **Sept 18–20, 1987:** Driftless Bioregional Network planning gathering *(page 14)*
- **Dec 3, 1987:** Pre-CROPP: Marketing Project Committee meetings *(page 31)*
- **Dec, 1987:** Definition of "organic" agreed upon by the marketing project committee. *(page 32)*

1988

National Context: May 12: Wisconsin Organic Crop Improvement Association is formed. Farm Bill passed without reforms. Drought

CROPP Cooperative:

- **Jan 5:** Name suggested by Spark Burmaster: CROPP (Coulee Region Organic Produce Pool) *(pages 31–32)*
- **Jan 18:** CROPP's first of four public meetings *(page 33)*
- **Feb 22:** 1st *CROPP Report* published.
- **March 10:** Incorporation papers submitted to State of Wisconsin. *(page 40)*
- **March 17:** Dairy group announces plans to produce 20,000 lbs of milk every other day. *(page 42)*
- **April 19:** Class B Preferred Stock issued. *(pages 62, 64, 68)*
- **June 15:** CROPP's first Annual Meeting elects first board. *(page 40)*
- **July 13:** First organic cheese produced by CROPP. *(page 44)*
- **Aug 10:** Membership approves by-laws.

Regions, Pools and Subsidiaries: Vegetable, dairy, meat pools formed. Wisconsin dairy pool

Number of Farms: 57 farms

1989

National Context: Feb 26: CBS airs *A is for Apple* on the Alar Scare.

CROPP Cooperative:

- Keith Johnson as first general manager *(page 54)*

- **Feb 22:** First refrigerated truck is purchased. *(page 66)*
- **August 1:** Purchase of La Farge Creamery building from AMPI *(pages 63–65)*

- **Sept 18:** Swiss Chateau Cheese Factory makes all CROPP's cheese.
- **Sept 25:** Cheese cutting and packaging equipment installed in La Farge. *(page 64)*

Number of Farms: 34 farms

1990

National Context: Organic Foods Production Act passed by Congress.

CROPP Cooperative:

- **Feb 23, 24:** First Wisconsin Organic Growers Conference sponsored by CROPP, OCIA and WWTC *(page 68)*
- **April 17:** Organic Valley brand is created. *(pages 61–62)*

ORGANIC VALLEY
FAMILY OF FARMS
CROPP COOPERATIVE

- **May 30:** First cheese (5,000 lb) with Organic Valley label shipped to California. *(pages 57–62)*

Number of Farms: 35 farms

1991

CROPP Cooperative:

- **Feb 5:** First management team formed. *(page 88)*
- **April 3:** First butter made by Westby Creamery. *(page 86)*

- **July:** First cultured butter is sampled by Board. *(pages 86–87)*

Number of Farms: 31 farms

1992

National Context: Rio Earth Summi

CROPP Cooperative:

- **Jan:** Equity program requires 5.5% of base for dairy pool. *(pages 65, 68)*
- **Fall:** Non-fat Dry Milk Grade A introduced.

Number of Farms: 34 farms

1993

Mad Cow cases peak in UK (11,000).

- George Siemon "gradually" becomes CEIEIO of CROPP. (pages 90–91)
- **March 2:** First Egg Pool Meeting (page 90)
- Egg Pool partners with Reedsburg Egg for packing. (page 66)
- **July 4:** First CROPP float in La Farge parade (pages 90, 92–93)

Egg pool formed.

49 farms

1994

- First milk carton design selected. (page 85)

- First fluid milk distributed in the Midwest. (page 98)

Midwest region

Minnesota and Iowa and dairy pools

84 farms

1995

- CROPP products enter first retail grocery chain. (page 123)
- Oregon Tilth Certified Organic becomes CROPP's official certification agency. (page 110)

Pacific Northwest region

Trout Lake, Washington, dairy pool

100 farms

1996

Genetically modified crops are introduced: corn, soybeans, canola and cotton seed oil.

- Brand gets a face lift. (page 144)
- Horizon offers to purchase the Organic Valley brand (page 127)
- CROPP proposes pasture language for National Organic Standards Board. (page 139)

Organic Meat Company LLC

119 farms

1997

- La Farge Appreciation Day sponsored by CROPP.
- Jim Wedeberg becomes dairy pool director. (page 95)
- First Kickapoo Getaway

Maine and McMinnville, Oregon, dairy pools

160 farms

1998

First organic restaurant in U.S.: Nora's in Washington D.C.

- CROPP purchases Chaseburg Creamery building. (page 125)
- Regional sales starts in the east. (page 116)
- Ultra pasteurized (UP) milk shipped. (pages 123–124)
- CROPP ships to Walmart and Target. (page 147)

WAL★MART

- First CROPP website is launched. (pages 178–179)

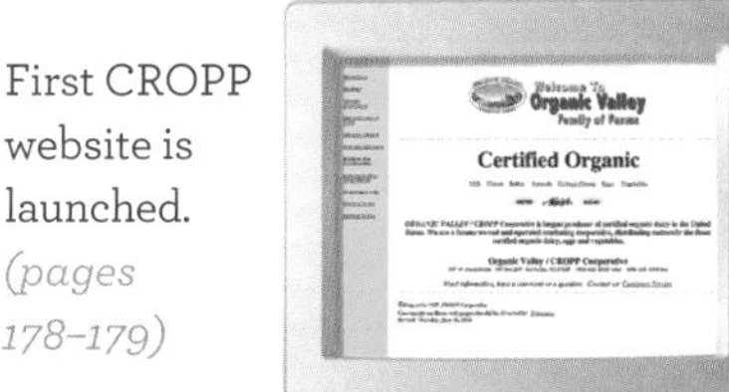

- CROPP celebrates 10th anniversary.

California and Lancaster County, Pennsylvania, dairy pools

182 farms

1999

Organic meat recognized by USDA.

- CROPP develops first in-house art department. (page 144)

Northeast region

Vermont and Myrtle Point, Oregon, dairy pools

244 farms

National Context | CROPP Cooperative | Regions, Pools and Subsidiaries | umber Farms

2000

National Context: National Organic Standards adopted.

CROPP Cooperative:

- OV Raw Sharp Cheddar wins American Cheese Society first place award. *(page 95)*
- Farm Friends Program created. *(page 128)*

Regions, Pools and Subsidiaries: Florida juice pool formed.

New York and Upper Pennsylvania dairy pools

340 farms

2001

National Context: Small Planet Institute founded by Frances Moore Lappé.

CROPP Cooperative:

- CROPP ships orange and grapefruit juice. *(page 142)*

- CROPP obtains $500,000 grant for meat brand. *(pages 141–143)*
- CROPP leads major letter writing campaign to urge USDA to pass National Organic Standards. *(page 139)*

440 farms

2002

National Context: USDA National Organic Standards officially enacted.

ORGANIC VALLEY

Family of Farms

CROPP Cooperative:

- CROPP enters its fluid milk into the "store brand" marketplace for the first time.

- Organic Valley creates "Pastures" brand regional cartons.

Regions, Pools and Subsidiaries: Ohio dairy pool

510 farms

2003

National Context: Cases of Mad Cow found in Canada.

CROPP Cooperative:

- The new-look Organic Valley brand begins to appear on packaging. *(page 144)*
- OV Salted Butter wins first place at World Dairy Expo.
- OV introduces single serve milk.
- OV Omega-3 Organic Eggs are introduced.

ORGANIC VALLEY Family of Farms SOY VANILLA

ORGANIC VALLEY Family of Farms SOY ORIGINAL

610 farms

2004

CROPP Cooperative:

- Organic Meat Company creates Organic Prairie brand. *(page 146)*
- Headquarters building built overlooking Kickapoo Valley. *(pages 151, 154)*

- Organic Valley SOY Flavors are introduced. *(page 148)*
- First Kickapoo Country Fair is held. *(page 182)*

Regions, Pools and Subsidiaries: Soy pool formed; Organic Logistics LLC formed.

Colorado and Western Washington dairy pools

680 farms

2005

National Context: US organic farm land doubles between 2002 and 2005.

CROPP Cooperative:

- "Dry Thursday": demand outstrips supply, CROPP drops Walmart. *(page 150)*
- Feed program started in anticipation of feed shortages. *(pages 142, 183)*
- CROPP holds first Staff Earth Dinner.

Regions, Pools and Subsidiaries: Texas and Indiana dairy pools

730 farms

2006

CROPP Cooperative:

- Farmers Advoca ing For Organic (FAFO) formed. *(page 145)*
- **July 29:** CROPP holds first annua stockholders meeting at Kickapoo Count Fair. *(page 182)*
- CROPP establishes its own pasturing standard. *(page 139)*
- Organic Valley becomes the #1 brand in the natural foods market.

Regions, Pools and Subsidiaries: Idaho/Utah, Kentucky and South Dakota dairy pools

917 farms

2007

- Chico State University and University of New Hampshire organic research herds join the CROPP dairy pool.
- First year of voluntary farmer-contributed fund: Farmers Advocating For Organic (FAFO) *(page 145)*
- CROPP Wellness Program created to encourage staff fitness. *(page 174)*
- Cashton Distribution Center opens. *(pages 183, 187–188)*

Tennessee, Virginia, Louisiana, Mississippi and Massachusetts dairy pools

1201 farms

2008

Global financial crash leads to massive recession.

- Roll-out of new products: OV Whipped Organic Butter and Vermont Cheddar Cheese
- Rideshare program for employees begins.
- Farmer Renewable Energy Program initiated. *(page 189)*
- CROPP celebrates 20th anniversary.

Grower pool formed.

New Mexico and Wyoming dairy pools

1341 farms

2009

Sputtering economy results in weak demand.

- CROPP has no growth; sales shrink by 1 percent. *(page 183)*
- First use of a dairy pool supply quota *(page 186)*
- Hiring freeze for employees *(page 183)*
- Partnership with Stonyfield involves licensing their organic fluid milk brand. *(pages 116, 123)*

1404 farms

2010

Florida freeze

H1N1 flu scare

- **August 1:** Quota system ends. *(page 186)*
- Hiring resumes.
- New products introduced: Organic Valley Feta Crumbles, Grated Parmesan and Drinkable Yogurt.
- OV orange juice supply dries up. *(page 138)*
- Tracking photovoltaic system installed at headquarters.

Missouri, Michigan, West Virginia and Connecticut dairy pools

1617 farms

2011

GMO alfalfa legislation approved by U.S. Government.

- Solar photovoltaic celled windows installed at headquarters to generate electricity. *(page 189)*

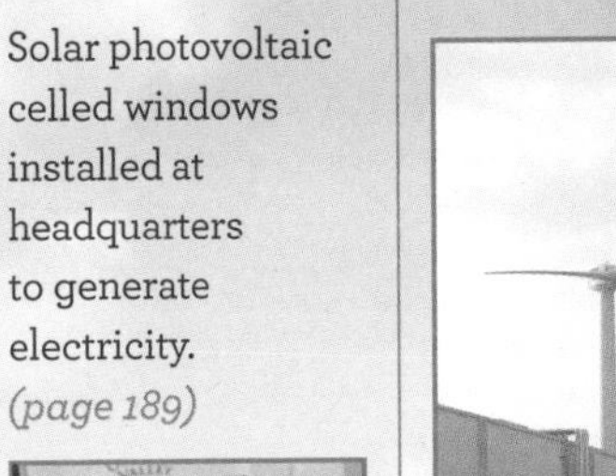

- Hot water and photovoltaic cell panels installed on the headquarters roof. *(page 189)*
- Organic Valley deepens its social media presence on Facebook and Twitter. *(page 179)*
- FAFO supports Anti-GMO Initiative with employees and farmers marching on Washington.

Australian beef farmers join beef pool.

1687 farms

2012

Severe drought devastates U.S. farms.

- **July 18:** Wind turbines dedicated at Cashton Distribution Center. *(page 188)*
- Quality, Research and Development lab constructed at headquarters. *(page 185)*
- Hillsboro Produce Consolidation Warehouse begins operating.
- American Singles rolled out as new product.
- Grassmilk pool created and special milk carton designed.

Northern California grassmilk pool

1814 farms as of December

2013

- 25th anniversary celebration includes the debut of *CROPP Cooperative Roots: The First 25 Years.* *(pages 1–204)*

- New-look logo and milk cartons developed for Organic Valley brand.

- Cashton Distribution Center addition to be completed in 2013. *(page 187)*

4. “A Rumbling Volcanc

"Poised to Explode"

991–1995

* Group Mind * Gibb's Triangle * Farmer-rur

* Role of the champion * Social fun *

"Back in the 1990s, there were no scientific studies to prove the benefits of organic food. Instead, there was intuitive knowledge." Mike Bedessem, CROPP chief financial officer

Mike Bedessem shed his conservative suit and wingtips after years in corporate America to grow organic apples in Gays Mills, Wisconsin. It was 1990.

"I was burned out," he says. He had been chief financial officer of a large, national company. The philosophy of organic and the chance to work with his retired father had major appeal. Bedessem was 12 years old when he first helped his dad with bookkeeping in the family accounting business. Now he and his brother, Kevin Dobbs, were tending Turkey Ridge Organic Orchard, growing Liberty and Macfree apples and introducing these unknown varieties to consumers in grocery store demos.

Only one year earlier in February, 1989, Ed Bradley of CBS TV's "60 Minutes" reported that "The most potent cancer-causing agent in our food supply is a substance sprayed on apples to keep them on the trees longer and make them look better." The substance was called Alar. Though a person would have to ingest excessive amounts of applesauce or apple juice to be endangered, the U.S. Environmental Protection Agency banned Alar's use in food production.

Suddenly, Bedessem recalls, wholesale distributors were calling Turkey Ridge Orchard and asking for semi-truck loads of organic apples. At best, Bedessem could offer only three pallets. "The Alar scare raised consumer awareness," he says, "but the retail food infrastructure decided that organic producers couldn't deliver." That was a challenge for the young organic industry, but the scare led regular folks to further question the safety of their food.

"People who chose the alternative lifestyle were early adopters of organic foods. It was a small market in the 1980s," Bedessem says. "By the time I was doing store demos in the early 1990s, there was evidence of an awakening among conventional, middle class consumers." Consider the wife of a Motorola engineer who Bedessem met at Fresh Fields natural foods store in Chicago. "She drove 45 minutes one way to buy organic produce because her husband suffered severe rashes if he ate conventional fruits and vegetables," he says. "People were beginning to see the link between allergies, illnesses and food additives. Stories like hers were telling. They made me realize there was something fundamentally wrong with our nation's food system and organic had a strong future."

← Monte and Laura Pearson's farm in Trout Lake, Washington

← Mike Bedessem was a CROPP board member before he joined the CROPP staff; many Amish farmers joined CROPP in the early 1990s.

CROPP's milk truck sported the Organic Valley logo in 1992. →

"This is our fifth annual meeting . . . . Sales are growing. We are in a dynamic, expanding business." CROPP Annual Meeting, March 21, 1992

"100 Farmers . . . Someday"

When Bedessem was introduced to the leaders of CROPP, the Cooperative had reached its first $1 million mark in gross revenues in 1991 ($1,128,257, to be exact). This was CROPP's first year without a net loss. The Cooperative's dairy farmers were making as much as $14.43 per hundredweight for their organic milk, about $3.40 above the conventional price. As demand for organic milk grew, CROPP would not have to sell as much of its milk on the conventional market at the lower price.

If they dreamed big, CROPP dairy pool members imagined they'd have "100 farmers . . . someday" and members of the pool were prepared to chip in money to help fund the additional costs of growth. CROPP had its own spiffy milk truck sporting the Organic Valley logo and a new "freezer on wheels" for cheese production, thanks to a grant from some generous Presbyterians. Later, the Presbyterians funded an upgrade of the La Farge headquarters electrical system. In those early years, CROPP could never be sure who their next benefactor would be; the cast of diverse lay and church supporters was broadly ecumenical.

Though CROPP's organic vegetable production—and sales—fell short of projections, many enthusiastic new Amish growers had joined the fold, too. Everyone had reason to be encouraged, because more consumers were buying organic, even though the U.S. was mired in a tough recession. The early 1990s was also an era when "downsizing"—the euphemism for lay-offs and job losses—hit many of America's once-proud companies.

Lack of access to capital plagued the young CROPP, but Jerome McGeorge told members and shareholders at the annual meeting in 1992 that the Cooperative finally had $135,000 in long-term financing—its first. To hearty applause, McGeorge said CROPP was only behind on paying two accounts. Perhaps the days of "creative financing," as McGeorge called it, would soon be over.

The first genetically modified crops were introduced commercially in 1996.

Though CROPP's hand-written annual report had a sort of human charm, this would also be the last one penned by McGeorge. CROPP was becoming the new agricultural force it set out to be and it better look a little more like it. "CROPP has established national leadership in organic dairy and organic standards," McGeorge declared. "We represent hope for the future of small and mid-size farms."

"I thought it was a lead pipe cinch that CROPP would get to $25 million in sales." Mike Bedessem in 1992

Mike Bedessem was intrigued and flattered when dairy pool member Dave Engel encouraged him to run for CROPP's Board of Directors. Bedessem's corporate experience with start-up ventures, early-stage financing, rapid growth, business systems and employee benefits would be a big help, Engel told him. The Board was composed of farmers with loads of common sense, but very little business experience. Bedessem joined the Board in 1992. Not long after that, he would resign and become CROPP's business manager, earning $11 an hour. Bedessem imagined he could still be a part-time organic farmer and business owner. "I thought I could do this

← Lila Marmel and her son, Dylan, in 1985

Arnie Trussoni and his wife, Tama, were the 10th organic dairy producers to join CROPP in 1992. Here, he is pictured with Justin. →

new job and two others at the same time, but it quickly became apparent that CROPP was going to take off."

No kidding. In just five years, from 1991 to 1995, the Cooperative's total sales would soar from $1.1 million to $9.1 million, and the total number of farm families depending on CROPP for their livelihood would reach 100 in 1995.

Even so, CROPP didn't look like a thriving enterprise when Dan Hazlett showed Bedessem to his office in the converted La Farge cheese factory that was CROPP's headquarters. Hazlett grabbed a pail on their way up the stairs. They stopped at a tiny corner office that Hazlett called "a rat hole."

"This is it," Hazlett said.

"Not too bad," Bedessem gamely said. "I've seen worse."

Hazlett set the pail next to Bedessem's desk.

"What's that for?"

"The leak," Hazlett answered. "You'll know . . . when it rains."

No Playbook to Follow

Lila Marmel, an early CROPP employee, organic vegetable grower and later a Board member, captured the early years of the Cooperative's steady—and sometimes stumbling—development in the *CROPP Report*, a quarterly newsletter. "The years 1991 and 1992 were really difficult," she said in her editor's note. "CROPP was a tiny organization and each step forward seemed like a phenomenal accomplishment. We were one of the lone voices, in one of the most economically depressed areas in our nation, trying to organize farmers and create a larger entity that could market organic products over a broader geographical area."

> **"We have succeeded as a poorly capitalized company in a unique marketplace with creative financing and the strong commitment of our members."** Lila Marmel, editor, *CROPP Report*, 1992

When opportunities came along for CROPP, there was no playbook to follow. Some decisions were controversial. It was 1992 when Natural Horizons, Inc. of Boulder, Colorado (better known as simply "Horizons," and later, after an official brand name change, "Horizon Organic" or "Horizon.") asked CROPP to sell them organic skim milk for their low-fat organic yogurt made in Madison, Wisconsin. The opportunity was a godsend. Skim milk was a by-product of making organic butter—as CROPP's butter sales grew, so did its skim milk inventory. Selling to Horizons meant they would receive the higher organic price and more organic dairy farmers could "get on the CROPP milk truck" and become members as the Cooperative's milk orders increased.

← CROPP produced its first Organic Valley ½ gallon milk cartons in August 1994; George Siemon and Spark Burmaster checked carton proofs.

General Manager Keith Johnson and Marcia Halligan at a 1989 board meeting →

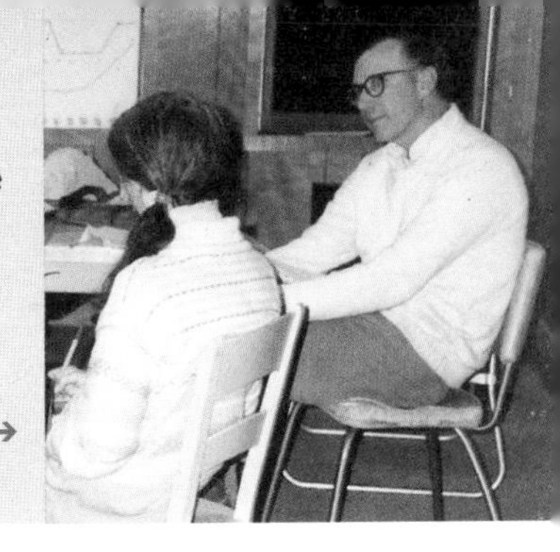

The decision was tough. "Horizons wasn't like us" mused Jim Kojola, CROPP's operations manager. "They were slick guys with Rolexes," he recalls. "We went back and forth on that decision to sell to Horizons. Are we helping our competitor? Is this helping us? Is it both? Can we really work with these guys? "

> "Delivering skim milk to Horizons for organic low-fat yogurt allows us to enter a broad market with a perishable product, which CROPP could not do with our own resources." *CROPP Report,* Winter, 1992

But let's be practical. CROPP needed the business to strengthen its bottom line and serve more organic farmers. Horizons was owned by two experienced businessmen: Mark Retzloff, owner of a chain of organic supermarkets, and Paul Repetto, a food company CEO with strong advertising experience. "Horizons had money, they had knowledge and—at the time—we knew we were going nowhere fast by ourselves," George Siemon remembers. "We saw our opportunity to sell to Horizons as a welcomed addition." That way, CROPP's dairy pool could focus on building its cheese, butter and new milk powder product (another good complement to CROPP's butter production). By selling to Horizons, CROPP was helping their future competitor become a national brand, and they were putting more organic products on retail food store shelves and into consumers' hands.

Very soon, Horizons introduced its new line of fluid organic milk to the Los Angeles market using milk that CROPP supplied them. Playing catch up, CROPP introduced its own branded fluid milk in Baltimore about a year later. CROPP would soon go head-to-head with Horizons nationally, and in just two more years, Horizons would make overtures to buy CROPP's Organic Valley brand.

> "Supplying Horizons got them launched nationally. It also got us to say, 'Hey, let's do it ourselves.' " Louise Hemstead, CROPP chief operating officer

Hunker Down and Survive

Just before CROPP made its decision to sell milk to Horizons, Keith Johnson, the Cooperative's first General Manager abruptly resigned in February, 1991, one day after a meeting of the CROPP vegetable pool. The members had struggled to meet their production goals and everyone was frustrated.

Johnson coordinated that pool and the overall business affairs of CROPP. He had served CROPP through its formative, high-stress years, working virtually nonstop. When money was tight—which it usually was—Johnson often took a pass on his paycheck. He was

This Butter Is Better

When his butter churn broke, Willi Lehner thought it was an omen.

The son of a Swiss cheesemaker, Lehner had learned to make butter the old-fashioned way. He experimented with stove-top butter-making and he carefully churned every ounce by hand. He made European-style cultured butter from milk produced by pastured Wisconsin cows. His butter was lush with live cultures and healthy Omega 3 acids. Lehner sold his butter at the Madison farmers market and—at peak production—he made 180 pounds in a single batch.

"The butter was unlike any other you could buy," says Lehner. "It was incredibly golden yellow." But it was hard to produce, and when Lehner's churn gave out, he took a six-month breather—that is, until George Siemon of CROPP Cooperative called on him in 1991.

CROPP had won a grant to market organic butter, but many ag experts scoffed. There was already an oversupply of butter because it had fallen out of favor with consumers who wanted to lower their fat intake. In fact, butter was selling for under a dollar a pound. Nevertheless, CROPP sought out Lehner. CROPP was already making organic cheese, so the next logical step was butter, another long-shelf-life product they could market nationally. Never mind what the margarine-lovers said.

"Some organic farmers were producing small batches of butter," Harriet Behar remembers, "so we decided to make unsalted cultured butter our first product. Maybe it sounds crazy, but we wanted to give people the best butter they could get. We asked Willi to be our consultant and the Westby Cooperative Creamery to make it." Westby already had the equipment, though the machine they used to wrap the butter in parchment and print on it dated back to the 1930s.

Westby's manager, Tom Gronemus and his staff needed plenty of coaching from Lehner and CROPP. For example, how could they capture the buttermilk and keep it free of dangerous bacteria before adding it back to the butter? (That would require lots of precision). How could they store the finished butter in a perfect, thoroughly frozen state before it was code-dated and shipped? (They did that in a freezer truck.)

Even with this steep learning curve, the Westby Creamery came through. "I was there when they turned out their first batch," says Lehner. "They made 1,500 pounds of beautiful butter in a short time, compared to my laborious process!"

When it was introduced, Organic Valley Cultured Organic Butter was the first produced in the U.S. commercially. (Over the years, Organic Valley's butters would win at least 25 national and international juried quality awards). CROPP's dogged insistence on the highest quality, combined with its ideal of innovation, raised eyebrows in the dairy universe, and orders were slow to materialize, at first.

Fortunately, Purity Farms, a leading producer of ghee (clarified butter), bought up half of CROPP's butter production. Jerome McGeorge remembers the call from Purity. "We can't believe it, we cook your butter and it smells sweet, not like commercial butter," Purity's man told McGeorge. "We're interested in up to 10,000 pounds. What's the price?"

When McGeorge said $2.05 a pound, twice the conventional price, he backed off. "A week later, Purity called to order 5,000 pounds," McGeorge says. "Not only did our butter smell sweet, but when they went to skim off the impurities that routinely rose to the top of heated butter, there were none. They sent us a bouquet of roses thanking us for selling to them!"

Organic Valley's packaging was simple and pure, just like the product, says Behar. "We had a sumptuous yellow label with a little blue line across the bottom that said, 'cultured.' It was simple and striking and our sales took off. The next big challenge was finding a use for all the skim milk produced in the butter-making process."

As CROPP found markets for its skim milk and butter production increased, the little plant at Westby was pushed beyond capacity. It was harder to keep a keen eye on quality and the inefficient production process sacrificed many pounds of valuable cream. Westby couldn't keep all the butter sufficiently cool, either. By 1998, CROPP would work a small financing miracle and buy a shuttered creamery in Chaseburg, Wisconsin, and convert it to a butter plant. In doing so, CROPP could produce more butter, faster, with even better quality and production controls.

Today, CROPP has four locations where its Organic Valley organic butters are made, but Chaseburg is the only place where the crème de la crème of butters—the cultured varieties—are made. "We bring fresh farm milk into Chaseburg and take the cream off the milk and make our cultured butter from that cream," says Louise Hemstead, CROPP's chief operating officer. "The milk is handled as carefully as milk can be handled. It's not moved from one plant to another. We take extra, extra care. Our European and Pasture cultured butters have 84 percent butterfat (Organic Valley's other butters are 80 percent butterfat). They have more Omega 3 fatty acids that are better for brain health, and they have more CLA (conjugated linoleic acid), an important antioxidant. They even have the Omega 6 fatty acids that are good for us."

It's enough to make Willi Lehner's mouth water.

a principled guy who worked with a cadre of idealistic, enthusiastic folks, many of whom were independent and not so easily managed. The term "herding cats" is apt.

Johnson was exhausted and disheartened and he said as much when he came to the La Farge office and picked up his Commodore 64 computer on his last day. Johnson found George Siemon and Harriet Behar at a café in nearby Westby that morning and broke the news. "Keith was high-minded and hardworking," says Siemon. "He deserves a great deal of credit for keeping us whole in those early years."

It was a difficult transition. "We had too many dreams; we were too idealistic," says Siemon. "When Keith left, we had to hunker down to learn and survive." Siemon was reluctant to step into the lead role at CROPP. As Dairy Pool Coordinator, he had his hands full and he quietly hoped that he could soon return to full-time farming.

It takes 10 pounds of milk to make a pound of cheese and 21 pounds of milk to make a pound of butter.

Johnson's departure was the catalyst that created CROPP's first management team including Siemon, Jerome McGeorge, Harriet Behar and Jim Kojola. They applied their natural, consensus-building style to making decisions as a team, and it worked. When CROPP started its dairy pool, for example, the original farmers plus Siemon met weekly to make decisions about the pool's pricing, function and future. They sat down together, talked through the facts and implications, gave everyone a chance to speak and made group decisions. Disagreement was a natural part of the process, but once a decision was made, each member's responsibility was to support it. It was similar to the American Indian tradition of elders seated in a "talking circle," each sharing their thoughts one by one and ultimately hearing the common thread of choice emerging.

This same "group mind" approach to making decisions is common at CROPP today. "A full, cooperative process may take longer to go through," says Siemon, "but in the end, you have built a foundation for the future and you have a wiser decision." Observers of CROPP's culture may say there are too many meetings to seek "group mind" and too many people in those meetings, but the process works and it started early in the Cooperative's life.

Relying on "Group Mind"

CROPP's governance structure depends on trusting this "group mind" and giving the farmer's voice priority. It has its roots in how Wisconsin's tobacco pools were run and CROPP's distrust of traditional cooperative practices. The Cooperative's Board of Directors, composed of organic farmers, is the final decision maker on such major items as policy, pay price, capital investment and operating budgets. As CROPP grew and added new "producer pools" of organic products (first produce and dairy, later eggs, meat, juice, soy and organic feed), pool executive committees composed of farmers were created to represent these organic products and—in the future—geographic regions, too. CROPP farmers elect members of pool executive committees for two-year terms, and the committees meet regularly to make decisions involving pool operations. They recommend their best thinking to the Board. It is highly unusual for the Board to act contrary to a pool executive committee recommendation.

"Our whole structure focuses on farmer-member engagement," says Jim Wedeberg, one of the original dairy pool farmers and, today, Dairy Pool Director. "We never wanted CROPP to become a cooperative that simply viewed our producers as numbers, and we never have."

> "CROPP's structure invites farmers to be heard at every level of decision-making. We rely on them to remind us of who we are."
> Jim Wedeberg, CROPP dairy pool director

Trusting Human Goodness

CROPP farmer-member Tom Frantzen introduced the idea of Gibb's Triangle (originated in 1978 by Dr. Jack R. Gibb), a concept that would shape CROPP's culture for years to come. "Gibb's Triangle compares two potential paths," says Siemon. "The first is built on prioritization that begins with trust in human goodness, then sharing common goals and communicating them effectively. This path minimizes rules and controls. In contrast, the other path relies heavily on rules and controls, while minimizing trust. Over the years, the Gibb's Triangle has kept our Cooperative viable. We must choose Gibb's first path and avoid the kind of tempting rigidity that wants to control everything that's human."

If CROPP's leadership is all about farmer involvement, then where does that put the paid staff, many of whom are farmers themselves? Is CROPP really "farmer run?" During the early years of CROPP, the lines between the Board of Directors and the staff were blurred. If Board Member Wayne Peters had a jackhammer, he would personally bring it to the La Farge building to break up an old cement floor. If CROPP needed a grant application written, then Jack Pfitsch, CROPP's Board President, would use his persuasive skills to write it. Today at CROPP it is understood that while the organization is "farmer-owned" and the Board (composed only of active farmers) makes the big, final decisions, the day-to-day affairs of the business are handled by

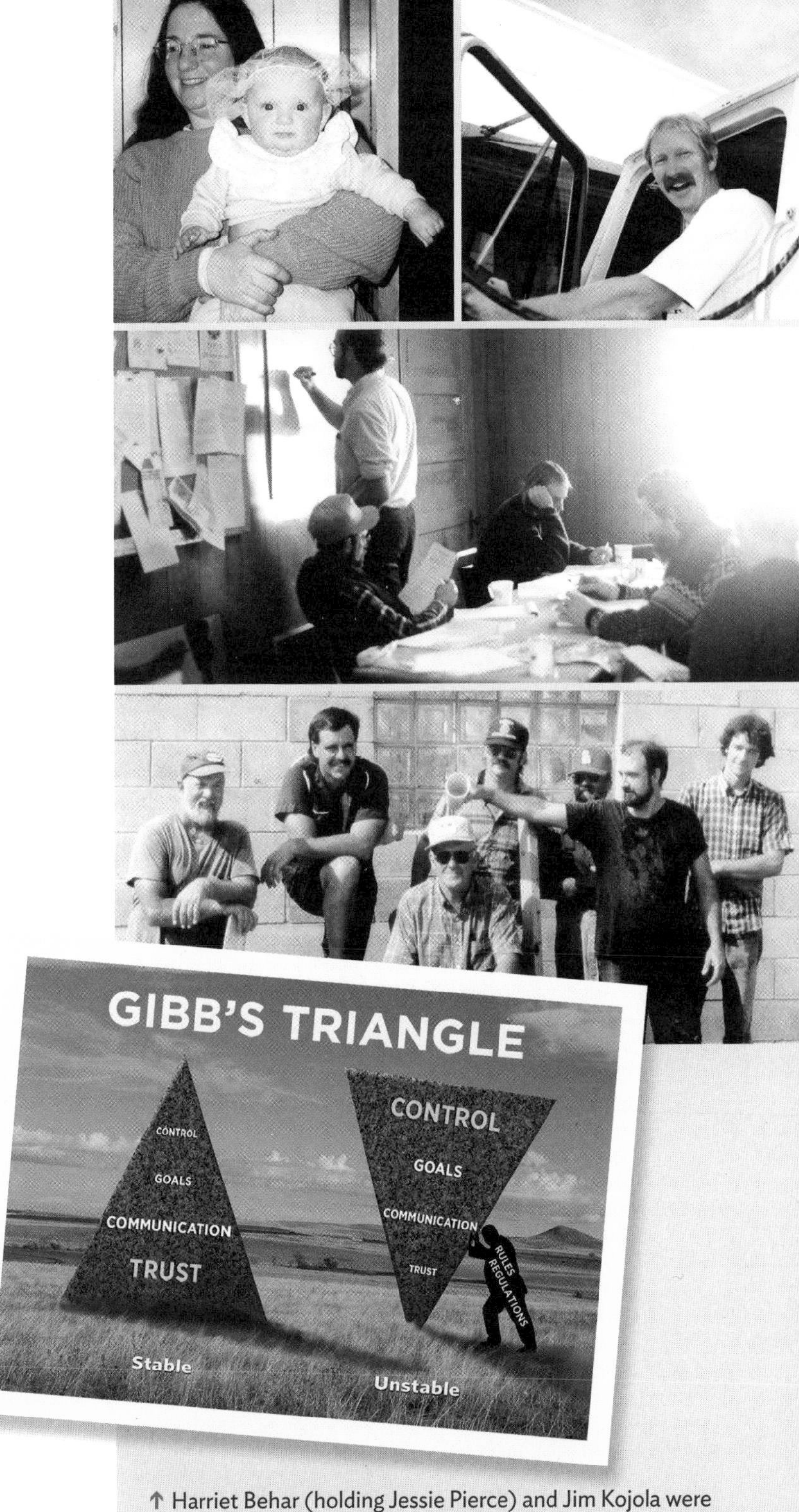

↑ Harriet Behar (holding Jessie Pierce) and Jim Kojola were members of CROPP's first management team; early CROPP planning meeting; CROPP staffers and farmer-members (from left: Dean Swenson, Carl Pulvermacher, Wayne Peters, Tom Forsythe, Jim Wedeberg, Dan Meyer and Bill Turner) combined forces to update their La Farge headquarters in 1996, and had fun doing it.

CROPP employees. Every important decision that the Board makes is informed by staff recommendations. "The staff has always provided strong leadership by preparing well-thought-out proposals that offer up all options so that the farmers can make the wisest decisions possible." Siemon says. "It is the employees who run the daily business."

CROPP is farmer-owned, you bet, but running the operation is shared by the farmers and employees today, as it was in the Cooperative's earliest days.

Not a Day Goes By . . .

George Siemon's personal decision was fast approaching. Would he return to full-time organic dairy farming or step fully into leading CROPP? Could he do both? By the end of 1992, the Cooperative's gross revenues had grown 80 percent over the previous year to $1.9 million, the strongest net profit so far. Revenues were projected at $2.6 million for 1993, another big financial leap. An ever-increasing number of farm families were joining CROPP, and the Cooperative was getting serious about more diversification: this time starting an organic egg pool, in 1993, that shipped about 5,000 dozen eggs each month.

The Cooperative had invited a business consultant in to advise them on its strategic future—not that they took every suggestion. "We have been challenged by business consultants to abandon diversity and focus on the one area in which CROPP is most lucrative," Siemon told annual meeting attendees in March, 1993. "We are obviously abandoning that advice."

> "We are facing our fastest expansion in CROPP's history. All of this is satisfying and that's good, but the headache of keeping it all going is sure real." George Siemon, *CROPP Report,* 1992–1993

The single, largest contributor to CROPP's growth, by far, was the organic dairy program that Siemon had coordinated from its origin. Not without good reason, the CROPP Board asked Siemon to become President of the Board of Directors in 1992 and assume the role of CEO in 1993.

Siemon thought back to 1987 when the idea of CROPP was being hatched. "Jane and I were cleaning squash and I remember talking about it," he says. "The real issue was that I didn't want to be a business person. I wanted to be a naturalist and spend my life in the outdoors."

Siemon talked with a close friend who had a cynical view of getting involved in CROPP's leadership. Maybe it was an ego trip, the friend suggested? "The organic philosophy believes that you can change the

← Clovis Siemon waits for his dad to leave the office after a long day in 1994. They'll head home in his 30-year-old Chevrolet Malibu.

Some COOs have brains *and* good looks: Louise Hemstead is Dairy Queen in 1995. →

world through collective action," Siemon says. "I didn't want to be a cynic. It was the cause, the excitement, the people and the energy that motivated me. I was in my mid-thirties. It was my 'prime time.' I was either going to find a meaningful mission . . . or not."

> "Not a day goes by when I don't look outside and say, 'Today would be a good day to . . .' I'm still a farmer."
> George Siemon

Siemon accepted the new leadership role, but he attempted to farm for another four years, only becoming more and more frustrated by his dual life. (Jane and George Siemon would sell their dairy cows in 1997 after 20 years of dairy farming.) Today, he feeds his naturalist yearnings with fishing, hiking, canoeing, bird watching and a little farming.

Sipping the Organic Tea

With Siemon in his new role, CROPP needed a Dairy Program Coordinator and the board's choice was 32-year-old Louise Hemstead, a quality assurance specialist for AMPI, then the nation's largest conventional dairy marketing cooperative. She had been a city kid whose dad bought a 160-acre farm in 1965 to introduce his children to the joys of nature. Her "dyed-in-the-wool hippie" parents grew their own vegetables and fruits and used the latest homeopathic remedies.

As young as age four, Hemstead spent weekends on the family farm (she even campaigned to live with neighbors so she could attend a rural school.) The love of agriculture captured Hemstead and she earned her college degree in animal science. But jobs were scarce when she graduated during the 1980s farm crisis.

Hemstead answered an ad for a job with CROPP in 1992. She met with the Board of Directors for three hours. "It was a hot August day in a room without air conditioning and I was wringing wet," she says. "They were so proud that they had made payroll for three months in a row and I was thinking, 'I have a steady job, I have health insurance, I have a paid vacation. I'd better keep that.' "

Hemstead offered to be a resource on dairy questions serving as a consultant free of charge. But not long afterward, when she discovered she would be traveling more for AMPI and her children were still so young, Hemstead decided to pursue the CROPP job. She took it on June 1, 1993, earning $11 per hour, a considerable pay cut from AMPI wages. CROPP's La Farge office was just seven miles down the road from the Hemstead farm, where Louise and her husband, David, milked 25 cows.

"The family farm mission of CROPP really mattered to me," she says. "My colleagues at AMPI came from family farms. Later, when I listened to George and Jim (Wedeberg) tell the CROPP story at farm meetings, it just seeped into my pores."

Employees at CROPP jokingly call the reason behind their crusader-like commitment to clean food and sustainable family farming "drinking the organic Kool-Aid." Hemstead prefers "sipping the organic tea." It took Louise and David ten years, but they converted their dairy farm to organic, too.

Our Customers Depend on Us

CROPP's new Dairy Pool Coordinator jumped right into her new role with a special meeting on quality.

"Hi There!" Hemstead wrote in the fall, 1993, *CROPP Report*. "Let's get together and have a good all-around discussion about quality and any problems that you may have on your farm. Please bring your spouses. Also bring any hired help who may be involved in your milking routine. In my opinion," she continued, "the

Finances Under a Full Moon

It's the rare Chief Financial Officer who recognizes the phases of the moon while balancing assets and liabilities, but Jerome McGeorge did. "As I write under the Full Moon," he told CROPP members and employees in the spring of 1992, "CROPP assets and incomes rise to record levels."

Finally there was enough profit to cover most of CROPP's pressing overhead costs and loans. Along with the hard numbers, the often-poetic McGeorge, in his report entitled *Dances with Organic Finances* (when the movie *Dances With Wolves* was popular), found reasons to use positive language—"financial progress," "apparent accomplishments," "major positive changes"—as he described CROPP's passage from what he called Stage 1 Start-Up Struggle to Stage 2 Sustained Growth. Those years were tough for CROPP, and McGeorge understood his role as both financial steward and "spiritual guide" of the fledgling enterprise and its people. "Spirits remain high," he assured everyone, "as we plan for the unknown future . . . ."

Holiday partiers Jerome McGeorge, Lee Kupersmith, Jennifer McGeorge and Jane Siemon at Thayer's Bar in La Farge, 1993

Spirits were indeed high at the March 21, 1992 annual meeting. Every detail of that annual meeting earned an exclamation point, proving this event was anything but ho-hum. "Lunch will be provided by the CROPP Staff and Board! Drawings for door prizes! Hobknob with organic producers! Vote on new Board members! Fun! Laughter! Excitement! Free Cheese! See you there!" Shoes optional.

The excitement of CROPP's mission and work helped build a "family of comrades" from the Cooperative's origin, says McGeorge. Employees and farmers shared social fun because their friendships extended well beyond the workday. From the Cooperative's start when many volunteers turned into employees, CROPPies (as CROPP employees are fondly known) saw their work as "more than a job." Farmer-members often pitched in to help with construction tasks at the La Farge headquarters, before and after long days on their own farms. It was a way to save money, to be sure, but the camaraderie built strong friendships, too. "It's OUR co-op," everyone agreed.

Social fun at CROPP became a binding agent. A convoy of CROPPies including employees, family and friends caravanned to Milwaukee in the summer of 1992 to take in a Brewers game, complete with a tailgating barbecue, autographs and sightings of baseball Hall of Famer, Rollie Fingers.

Another gang of CROPPies borrowed a hay wagon and constructed a farm scene, with hay bales, pint-sized

cows and pint-sized offspring of employees and farmer-members. "Family Farms for the Future," the Fourth of July float declared, as it rolled down the La Farge main street.

CROPP's Christmas party welcomed employees, farmer-members and friends to Fritz and Helen Gudgeon's home for dinner and a surprise talent show.

"Enter the National Mooing Contest," Lila Marmel, editor of the CROPP Report goaded readers. "It is my earnest opinion that organic farmers should be represented in this

competition! Wow 'em with your most authentic moo (no real cows allowed)."

Ping-pong, soft ball, euchre card games and canoe trips on the Kickapoo River were all part of CROPPie

Farm Aid, September 2003, Columbus, Ohio

fun, often lubricated with some of Wisconsin's finest beer.

Proving that a fun social outing could include learning, CROPP's farmers built their own traditions with farm tours, pasture clubs and regional meeting picnics. The Hyde family hosted a "Twilight Tour and Seed Swap," and the Zepczyk Farm invited fellow CROPP farmers to share ideas about pasture renovation. Mike Bedessem's Turkey Ridge orchard hosted an organic orchard field day, and the Wedeberg family offered a field day focused on rotational grazing. Dave and Marta Engel welcomed farmers to their farm where Marta, a practicing vet, described homeopathic and herbal tincture treatments for dairy cows, an effective alternative to antibiotics (yes, tincture of garlic and vodka does work). At each of these events and many more over the years, CROPP's growing membership of organic farmers became not only more effective producers, but better friends, too.

21st Annual Vernon County
Dairy Breakfast
Saturday June 6
A breakfast buffet will be served from 6 a.m. to 10 a.m.
Peters Farm, Chaseburg
Hosts: Wayne, Roger & Rory Peters
SERVING: Scrambled Eggs, Sausage, Pancakes, Orange Juice, Coffee and a Full Complement of Delicious Dairy Products
COST: Adults $3.00 Children - 10 & Under $1.00
THE PETERS FAMILY
The gala event will feature local entertainment and a program featuring the Vernon County Dairy Queen and princess, local politicians, royalty and celebrities.
MY MOTHER EXAMINING THE COW MANURE BETWEEN HER TOES. SHE SEEMED TO FIND ALL THE FRESH PATTIES, BUT HAD A GREAT TIME!
Vermont Getaway

← Spring brings more milk as calves are born and happy cows graze on fresh grass and clover.

Jane Siemon is every inch a farmer. →

best way for us to grow is to consider ourselves the right hand of the plant that manufactures our products. We can be proud of our high quality standards, and our customers depend on us for it."

With Hemstead's strength in quality assurance, one of the first things she did was check all of CROPP's Organic Valley cheese. "We had strict antibiotic and coliform testing programs at AMPI," she says, "because we made cheese for pizza manufacturers. E-coli wasn't the household word that it is today. I tested CROPP's cheese, and about 35 percent had E-coli in it." Hemstead discovered this problem close to the weekend and CROPP didn't have a sufficient reserve stock.

"I told George, Harriet (Behar) and Jim (Kojola) that we had to get rid of the cheese." The newly-hired Hemstead summoned up her resolve and quietly said, "Either the cheese goes or I go."

Hemstead drove home in tears. She thought she might be out of a job.

CROPP got rid of the cheese and found a new cheesemaker. The incident demonstrated CROPP's values in action, she says. "They were committed to providing a healthy food source that was free from chemical contaminants, as well as free of bacteria and other contaminants. That has been our mainstay value all along."

Organic cheese magic would begin to take shape in 1994, when Louise Hemstead persuaded Phil Vantatenhove, master cheesemaker of Gibbsville (Wis) Cheese, to certify his fifth-generation family cheese plant for organic production and make Organic Valley's Raw Sharp Cheddar. By 2005 a *Best of Show* was in hand from the prestigious U.S. Championship Cheese Contest—along with a Bronze medal that same year from the World Cheese Awards in London.

> "Our Cooperative feels like a rumbling volcano, poised to explode with sales growth as we enter spring flush." George Siemon at the CROPP Annual Meeting, spring 1994

A Double Flush

Every spring, dairy families can expect a "spring flush" of more milk, as cows give birth to calves and graze on luscious, new grass and clover in their pastures. But more than the dairy cows were flush in the spring of 1994 when CROPP's farm families reviewed their Cooperative's performance.

CROPP was flush with a 100 percent increase in profits (as small as they were) in 1993 over 1992. The Cooperative was starting to see a steady upswing in demand for organic dairy. In previous years, CROPP had sold a hefty percentage of its organic milk at lower conventional prices because organic demand wasn't yet strong enough. That was a necessary short-term decision, but not to be continued for long. The year 1993 signaled the start of a positive pattern: all of CROPP's skim

← Travis Pearson, milking one of the Jersey cows on his family's farm in Trout Lake, Washington

CROPP's milk truck; Wayne Peters (second from left) hosts a gathering on his farm in Chaseburg, Wisconsin. →

milk sold at organic prices and the increased demand for CROPP's organic butter justified a higher price, too.

The marketplace was saying that CROPP would need more dairy farmer-members—and soon. But inviting more farmers to get "on the CROPP truck" wasn't as simple as shouting a hearty "welcome aboard!" The Cooperative would have to secure additional financing and certification by the State of Wisconsin as a licensed dairy plant.

> "We are finally getting some business financing. The NFO helped to finance us, though they never wanted to be our banker."
> CROPP Annual Meeting, February 1995

Dairy plant licensing was the easier of the two tasks. To accomplish the financing, CROPP had to leave the NFO's secure financial nest. That meant that CROPP would finally own its cheese inventory and receivables that the NFO had carried on its books and financed for years. In a generous arrangement, CROPP bought its inventory from the NFO by repaying a $223,000 loan over four years. Each farmer-member contributed to paying off that debt with 25 cents per hundredweight of milk that each farmer's herd produced. While it was a costly outlay, this meant that CROPP would qualify for longer-term operating loans that are traditionally based on "collectable receivables."

Milk Coming Out of Our Ears!

With growth in demand, CROPP needed to manage its milk supply so it didn't exceed or fall short of what the market wanted. This would become a top priority in 1994 when CROPP created a system to track farmer production, holding everyone accountable for the first time. The system would never be perfect. Balancing supply, demand and utilization—especially for CROPP's organic milk—would become a perpetual challenge as the Cooperative continued to grow in numbers of farmers, customers and geographic reach. Ensuring that the right amount of organic milk was produced and paid for at the fairest price, turning that milk into organic products, and delivering the products to the right place at the right time would be a high wire act that only a few CROPP dairy nerds with calculators, maps and push pins could adequately explain.

> "We manage the supply to meet the demand versus letting the demand drive the price."
> Louise Hemstead

The "supply management" part of this equation had its origin in a surprise, says Jim Wedeberg, a dairy farmer and CROPP's dairy pool director. "We started 1994 with 24 organic dairy farmers and we finished the year with 48," he says. "Our sales projections, especially for organic butter, showed that we would need more

View from the Milk Truck

Clarence Rollins remembers his grandfather farming with horses on the family spread just east of Vernon County, Wisconsin, and hauling cream to the local co-op for processing. Everyone farmed the organic way in those days, he says, because there was no other option.

Today, Rollins, a former dairy farmer himself, drives a 6,000 gallon Organic Valley 16-wheeler milk truck up and down the rolling hills of Wisconsin's Driftless region. He picks up milk at 17 organic dairy farms on a 134-mile route and it takes him about 11 hours. His route begins about 6:30 a.m. "It's a long day," he says, "and I'm Organic Valley's oldest driver, but I've farmed all my life and worked road construction. I'm used to it."

On his route, Rollins is treated to multi-colored sunrises, glimpses of wildlife and the natural world at its best. A Wisconsin native, he knows how to tackle the ice and blizzards of winter.

The farmers on his route are "just like family," Rollins says. "I get well-acquainted with them." He pulls into his first stop, the Sand Hill Organic Farm, member number 946 of CROPP Cooperative. The sun is rising over the farmstead where Gary and Rhonda Parr milk about 35 Jersey cows. The family dog barks a greeting. It is Rollins' job to measure the milk's weight (2,577 pounds for this load represents four milkings), record the milk's temperature (there is an acceptable range of degrees for safe transport), and load the milk according to organic sanitary handling standards. Everything he does is recorded on a daily log.

Rollins admires the hard work and discipline of organic farmers. "It requires more labor than conventional farming," he says, "but that effort pays off in greater knowledge about how to care for the land and the livestock. If you were a sloppy organic farmer, you wouldn't stay in business for long."

He is amazed by some of the creative practices that organic farmers use to raise crops and livestock without chemicals. Rollins is especially impressed by the commitment of young organic farmers. "One of them told me, 'I made up my mind I was going to go organic. I knew I had to make big changes, but when I did, I was confident I could feed my cattle, take care of my land and farm it successfully.'"

"If you ask me," says Rollins, "those organic farmers are very good caretakers of the land."

milk." CROPP barely kept up with its milk orders in early 1994, so George Siemon and Wedeberg got busy recruiting more organic dairy farmers in Black River Falls and eastern Wisconsin to meet the expected, higher demand. "All of a sudden," Wedeberg says, "we had milk coming out of our ears! We had 20 percent too much. How did we miscalculate?"

As it turned out, some of CROPP's existing 24 dairy members had expanded their herds without the Cooperative's knowledge. "Our dairy executive committee decided that we needed to manage our milk supply, so we came up with our 'base plan.'" CROPP was doing something revolutionary in the dairy industry, a practice successfully implemented years before by tobacco cooperatives: they were managing their milk supply to meet market demand, rather than letting demand determine their pay price. In the conventional milk market, oversupply meant lower prices, and the free market set those prices.

A cow udder holds between 25 and 50 pounds of milk.

Beginning in 1994, each farmer's "base" at CROPP was calculated by looking at that farm's previous three years of production from August through July. The average over those three years became the farmer's base. If a farmer produced more milk than his base, CROPP would pay the lesser conventional milk price for the overage. There was no penalty if a farmer produced less. "In order to maintain a stable pay price for our farmers, we learned that we had to effectively monitor how much our members were producing," says Wedeberg. "It was a very controversial decision at first. We didn't want to control our farmers' lives. We just needed insight into production so we could make better decisions."

"Up to May 1, 1994 we have paid our farmers $15.50 per hundredweight with an increase in May to $16.10," George Siemon reported in the summer of 1994. "We are planning another increase to $16.50 in July, striving to get closer to our ideal sustainable pay price of $17.50 (they would make it in 1997). Our farmers have been patient and supportive and we need to implement these raises to reward them adequately." During that same time in American agriculture, the conventional milk price hovered around $12.00 per hundredweight.

With time, tweaking and more challenges of oversupply and undersupply, CROPP would arrive at a milk management system that the Cooperative today calls "active base." Production is now tracked and calculated monthly instead of annually, and a farmer's milk production, pay price and Class B stock in CROPP (based on the farmer's production and price) are automatically re-calculated every February.

"The most common complaint I hear about our eggs is, 'Why can't I get more?'"
David Bruce, *CROPP Report,* 1994

Here's to Happy Hens

Each spike in organic demand and sales seemed to make the operation of CROPP more complex. The Cooperative posted gross sales of $5.6 million in 1994, compared to $2.6 million in 1993. Alongside CROPP's robust growth in dairy, the vegetable pool multiplied its production by more than two times in 1994 and CROPP sold all the organic eggs that their happy hens could produce. Mike Bedessem, who eventually succeeded Jerome McGeorge as CROPP's chief financial officer, reported that CROPP could easily expect sales reaching $9 million in 1995—and that's not chicken feed.

From the start, CROPP farmers saw organic egg production as a valuable second source of income when,

↑ CROPP sold all the eggs its farmer-members could produce in 1994; Steve and Bev O'Reilly (along with Steve's brother, Dave) were CROPP's first farmer-members in Minnesota. Their first organic milk load in 1994 was processed by Schroeder Milk Company in St. Paul. Jim Kojola loaded boxes of that first shipment.

let's say, drought or floods damaged vegetables or predictable seasonal variations in milk supply reduced dairy income. Organic Valley eggs in the grocery dairy case also reminded consumers of the brand as they walked the food aisles. Egg sales increased and so did other branded dairy products. In time, organic egg production was so attractive that some of CROPP's farmer-members produced eggs exclusively. "The girls" lived well, with time in the pasture, three times more space inside than in conventional operations, and all the other animal benefits of being raised organic.

With all this growth, it was about time CROPP owned more than one computer, so Mike Bedessem asked the board for ten more. "Mike told us we could run CROPP forever if we bought ten computers," Wayne Peters, dairy farmer and Board chair recalls. "How about this, Wayne," Jerome McGeorge interjected, "You and I can split a computer." To that, Peters quickly answered, "... and I'll bring the splitting maul!" (Peters and McGeorge still manage to live full lives without PCs.) Many years later, as CROPP's computing and information technologies (IT) expanded exponentially, George Siemon gave new meaning to the acronym IT on a particularly vexing day. "IT," he said, "stands for In Trouble."

Crossing the Mississippi

Steve and Dave O'Reilly had been organic dairy farmers since 1973 on the family homestead in Goodhue, Minnesota, about 50 miles southeast of the Twin Cities. By the early 1990s, the growth of consumer interest in organic prompted the O'Reillys to ask CROPP what it would take to organize a new organic co-op in Minnesota. "CROPP was the only dairy opportunity available in the Midwest, unless a group of farmers decided to do it themselves," Jim Wedeberg remembers. The O'Reillys traveled to La Farge to learn from CROPP. It wasn't long, says Wedeberg, "before they called us and said, 'Why

Gary Lohmeyer of Goodhue, Minnesota, was one of CROPP's first farmer-members when the Cooperative ventured across the Mississippi River. →

don't you just come across the Mississippi and get our milk, instead of us starting our own thing?'"

Why indeed?

"Crossing the Mississippi to Minnesota was big for us," Louise Hemstead, CROPP's COO recalls. "Were we willing to move outside our comfort zone? What if we couldn't bottle all the milk in Minnesota and we'd have to haul some of it back to Wisconsin?" It seemed like a big risk.

Though CROPP's farmer-leaders knew all too well the risks of farming, they were cautiously conservative in making decisions involving their members' livelihoods. "I've heard people say we don't have a high tolerance for risk," says Hemstead, "but let's talk about getting into cheese; let's talk about deciding to sell fluid milk. When it comes to helping organic family farmers, I think we have a high tolerance for risk."

"A stable price and a good market means we can stay in farming and be successful. We can give our kids the opportunity to stay and farm, too." Steve O'Reilly

CROPP crossed the river to Minnesota. Brothers Dave and Steve O'Reilly had their first milk load pick-up on October 4, 1994, and it was processed by Schroeder Milk Company in St. Paul. That load was part of CROPP's introduction of Organic Valley fluid milk nationally. CROPP's new Minnesota pool of three farm families produced 35,000 pounds of milk every other day. In that same year, CROPP also crossed the Mississippi to pick up milk from its first farmer-member in Iowa, Mark Kruse. In its characteristically conservative way, CROPP told two more eager Iowa farmers they'd have to wait a while until the Cooperative was sure it had enough sales volume to welcome them.

"The best cooperative members are people who have considered creating a co-op themselves." Jim Wedeberg

"We're On Our Way"

It looked like CROPP was destined to expand beyond its home state. But just as this new image of CROPP as a Midwestern cooperative was settling in, the Pacific Northwest beckoned.

Geographic growth would test CROPP's cooperative spirit, George Siemon said in the summer of 1994. "I'm studying other bigger federated cooperatives to learn what—and probably mostly—what *not* to do! When we do grow into new areas, we will have a dairy executive committee with local regional pools having their own meetings. We set out to get something done with CROPP and we can certainly say we're on our way."

CROPP's geographic expansion to 1,814 member farms in 35 states, two Canadian provinces and Australia by October 2012 either started with farmer requests like the O'Reillys or new customer opportunities in often distant locations.

Cascadian Farm of Washington State was one of them. Gene Kahn, a 24-year-old grad school dropout from Chicago wanted to make a difference in the world, so he chose organic farming near the Cascade Mountains. Cascadian went on to become a major producer of organic packaged foods with Kahn as its CEO.

In 1994, when Kahn wanted to make an organic ice cream bar, he asked CROPP if they could supply the cream. All CROPP needed were organic dairy farmers in the Pacific Northwest.

The Force of Nature

To help make that happen, Kahn introduced CROPP to a force of nature named Theresa Marquez. The granddaughter of a Michigan dairy farmer, Marquez grew up watching her mother tend their family garden using manure, compost and no poisons. With a family of eight children, food preparation and cooking from scratch were a daily event. Mom was well aware that healthy children and healthy food were linked—meals were carefully considered and the family always ate together.

Right out of college, Marquez—a self-described 1960s hippie—started her career in the Pacific Northwest on the cutting edge of natural foods store management and merchandising. A small natural foods cooperative in Portland—Food Front, whose idealistic slogan promised, "Food for people, not for profits"—was the training ground. There were few organic selections for consumers in the 1970s and 1980s, mostly produce and grains. Organic was born out of the natural foods industry, Marquez says, and from the work of natural foods advocates who emphasized clean foods and ingredients without preservatives.

Natural foods expos on both coasts were hotspots for clean food pioneers, advocates and entrepreneurs. It was at Expo East in Baltimore that Marquez ambled past Gene Kahn of Cascadian Farm talking with George Siemon about starting an organic dairy program. Kahn pointed at Marquez and turned to Siemon. "Hire her! She knows everybody in the business in the Pacific Northwest. She knows the farmers and the markets and she can get it done."

Soon afterward, Marquez sat down with Siemon. "Do you think we can sell 2,000 cases of organic milk in the Pacific Northwest?" he asked.

"Slam dunk," she answered. "You're going to need a lot more than 2,000 cases."

"Okay. Write me a proposal and we'll present it to the CROPP Board." The Board unanimously agreed to set Marquez loose on their behalf to sell CROPP to retailers

↑ Theresa Marquez and Andy Radtke, graphic artist, in 1999; Monty and Laura Pearson on their Trout Lake, Washington, farm have the gift of music; George and Jane Siemon toured farms in the Pacific Northwest in 1994 as the Cooperative expanded to that region.

and help recruit organic farmers and markets in the Pacific Northwest.

Marquez's role and those of CROPP's first Pacific Northwest farmer-members exemplifies the importance of champions in the Cooperative's growth. Despite CROPP's collective roots and team management style, the founders learned early about the importance of having a champion to lead a new initiative. Someone needs to be the point person to make a group process work and Marquez was the champion in Washington and Oregon. It is equally true that the Pacific Northwest farmers who joined CROPP were organic champions. They staked their livelihoods on organic farming and proved to others that it was a sound choice.

Selling to natural foods retailers was easy by 1994, Marquez says, because consumer demand was clearly growing, but converting conventional farmers to organic was "tough as nails. Would you commit your whole livelihood to a product that might not have a market?" Marquez says. "In time, the more successes we had in the market, the easier it was to sell to farmers. Our first members were pioneers because they largely farmed the organic way. You could either call them 'progressive' or 'old fashioned.'"

"I Closed My Eyes . . . and Jumped"

Monte Pearson was one of them. He and his wife, Laura, started dairy farming on his grandfather's 87-year-old farm near the Cascade Range in tiny Trout Lake, Washington, in 1970. They milked about 100 Jersey cows and, he says, "We basically farmed organically because it was our mindset and our philosophy." When ag experts advocated a synthetic growth hormone called BST to boost milk production, Pearson rejected it. "We didn't want to stress our cows that way," he says. "We looked for a way to market our BST-free milk."

That led to a meeting at Theresa Marquez's home in Portland, Oregon, where several farmers gathered to meet George Siemon. "We're conservative people out here," Pearson says, "and George wasn't a conventional looking guy: long hair, Birkenstocks, an adult hippie. We hadn't met Jim Wedeberg and Wayne Peters who looked and sounded more like us."

Pearson and other Pacific Northwest farmers were skeptical of what organic represented. "I saw it as an elitist movement of well-educated, upper middle class people." Pearson met with Siemon and Marquez several times and he asked his local cooperative, Darigold, if they would allow him to hold a second membership in CROPP. "I didn't want to give up my membership in case this organic thing didn't pay off," Pearson says. The co-op said yes, but a board member took Pearson aside and warned him, "You can't farm without antibiotics. What are you thinking?"

Pearson wasn't all that certain of CROPP's future, but he went ahead and his farm qualified for certification in nine months. "I closed my eyes, plugged my nose, and jumped," he says.

That was 1994. In 2012, Pearson and his son, Travis, in his mid-thirties, have plans to expand their herd to as many as 250 cows and possibly buy another farm. When Pearson got on the CROPP truck with his first milk pick-up in August, 1995, there were 11 conventional dairy farmers in Trout Lake, Washington. The remaining three farms in the valley today are all organic.

> "We are no longer trucking fluid milk to the Pacific Northwest. Our Northwest pool is now fully certified." Jim Kojola

The True Heart of a CROPPie * *Wanda Lewison*

When I first started working at CROPP 21 years ago, I knew that if any small business was going to make it, CROPP would, because we were really together, and we were selling organic food. There was new awareness about the goodness of organic food, and this awareness has grown over time along with the business.

I grew up on a farm here in Vernon County, Wisconsin. We had around 10 to 12 cows we milked by hand, ran about 40 head of beef cattle and raised seven acres of tobacco. In addition to farming, my dad was also an over-the-road truck driver. He saw the way they sprayed the potato and broccoli fields so heavily with chemicals that the ground would be pure white, so he made the decision not to use any sprays on our farm. This meant we spent the summers doing the hard work of cultivating corn and, of course, hoeing tobacco—as kids, we always wished we could have sprayed instead of having to hoe, but as a kid you don't get much say in anything.

We didn't call it organic; us kids always complained that we had to farm the old fashioned way. I didn't really think of it as farming organically until I was hired at CROPP. I guess dad knew best!

I believe what really helped shape CROPP in the early years was George Siemon saying to us, "The business is like a wheel; every one of us is a spoke in that wheel, and every spoke is equally important to the wheel, and the better the business does the better we are all going to do." I believe this message is what made us all feel that working at CROPP was more than just a job. We all joined together and did whatever it took to grow the business and make it successful. We all grew very close while doing so.

When CROPP occasionally hired outside consultants to give us advice on how to keep up with the growth of the business, it made me realize how the feel of CROPP was so different. Even though someone from the outside may have been able to give good business advice, the thing that they could never know—because it wasn't anything you could see—was the spirit of CROPP, the heartbeat, and that came from all of us employees and farmers who really worked for and with each other.

I feel that CROPP still has this great personality: kind, non-judgmental, open-minded, loyal, very caring. We are all still "spokes in that wheel."

Now that I have many years invested in this cooperative, I can't picture anyplace else I would rather be working—other than ranching out West somewhere, of course. I'm still a farmer at heart, and working at CROPP has allowed me to hold on to my heart.

{ CROPP financial analyst Wanda Lewison is a lifetime resident of Vernon County, Wisconsin, where she nurtures her passion for horses and riding. She is a mother of three beautiful daughters and "Namma" to two of the greatest grandkids in the whole world. }

The Organic Path from Consumer to "Citizen-Eater" * *Theresa Marquez*

Harriet Behar, CROPP's first marketing director, penned our first tagline in 1990: "Farmers and Consumers Working Together for a Healthy Earth."

CROPP was marketing to what was then the largest and most lucrative demographic group: the post World War II Baby Boomers. The first wave of the Boomers graduated from high school in 1964 and I was one of them. Some Boomers started an "alternative food movement" because they had a strong desire to rethink our relationship with food and farmers. Retail cooperatives and natural food stores sprung up across the country, often in garages, small warehouses or church basements. Shoppers at these new-format stores were seeking food without preservatives, additives or artificial ingredients. In this atmosphere, the organic "movement" was born, and I was lucky to be one of the first organic advocates and consumers.

At our home, we subscribed to Rodale's Organic Gardening *and had a big garden. We shopped in bulk. We refused to vaccinate our kids. Our doctor was a naturopath. TV was taboo in our house and we did not use a garbage disposal, but we had an active and revered compost pile. We didn't just want different food, we wanted a different set of values, and organic became a critical part of a new lifestyle.*

A key factor in the growth of the organic industry was the use by the conventional dairy industry of recombinant bovine growth hormone (rBGH, also called BST), which, when injected into cows, increased a cow's milk. Many of us were outraged that this was introduced into a sacred food for our children, milk. And we were outraged at how it abused the cows. Our distrust for corporate food and chemical companies grew.

Today, the organic consumer segment is dominated by the educated female. An estimated 70% of new organic consumers are moms with young kids, concerned about what is clearly a broken food system. Whereas in the 1950s and 1960s childhood cancer was unheard of, not so today. The obesity crisis, the diabetes crisis, cancer, autism, ADHD, reproductive problems and asthma are all too common and increasing each year. We are now the fattest people in the world and the children born today will not live as long as their parents. These health woes worry moms who just want healthy and happy children. They embrace the Precautionary Principle.

We now recognize that the consolidation of our food and agricultural sectors were not—and are not—in the best interest of consumer health. This consolidation has birthed our cheap food system. Where early adopters of organic were motivated by respect for the environment, this has taken a back seat to personal health. The notion that our food is making us sick—and the reverse, that food can heal—has become widespread. Today these concerns contribute to the robust growth of organic foods. For people who want help in changing their behavior and culture concerning food, organic offers a step in the right direction. These consumers of organic are aware that their health prob-

lems—their own or someone dear to them—are directly caused by the food they eat. Perhaps they are fighting cancer or are diabetic or overweight. Many people with health issues are yearning to change their eating habits. Organic food, the organic lifestyle, *is a lifeline and a positive step in the right direction.*

The organic consumers of today have evolved into better educated, more sophisticated consumers. In fact, some of us don't even like being called "consumers." We "core" organic advocates are "citizen-eaters." We know that we vote with our forks and with our food purchases. We citizen-eaters continue to be passionate about recycling, gardening, exercise, living lightly on the earth, cooperation.

And let us not forget the subjects of taste and quality. After all, this is food we're talking about: a sensory experience of taste, smell and sight. A number of organic citizen-eaters are "foodies"—passionate about all aspects of food and redefining quality every day. We Americans want it all—taste, quality, health, environmental benefits and fun! Organic Valley and Organic Prairie products deliver on all of these attributes.

No matter what portal new entrants to organic come through, they are constantly evolving and changing. Once a consumer becomes interested in organic, they become exposed to a rich culture of passionate food activists and food lovers committed to the Good Food Movement. Hopefully they learn that we are all connected and we really are what we eat. We want to be good moms, good earth stewards, good cooks and good companions at the dinner table. We must have food to live, but we also find great enjoyment in it.

For CROPP, our food and the way it is produced is all about hope—hope for change for the good, hope for food for everyone, hope for healthy children, hope for a clean environment, hope for deepening our connection to all of life.

{ A forever foodie and long-time marketing director for Organic Valley, Theresa Marquez, in semi-retirement, continues her tireless work as CROPP's mission executive. For more than three decades, Marquez has been a driving force in the national organic industry. }

5. “Organic Footprint”

Spans the Country

FARMY
* The 3-legged stool * Build the business

ther the buildings * Partnerships * Compromises *

"A dairyman can't be stupid, *but* if he has a little craziness in him, he's more likely to make it." Doug Sinko, CROPP dairy pool coordinator, western region

Neighbors thought Doug and Sharon Sinko were more than a little crazy when they became the first certified organic dairy farmers in Oregon. A third-generation farmer and high school teacher, Sinko converted from a conventional dairy to organic in 1993.

But CROPP didn't think Sinko was crazy. By the mid-1990s, the West Coast was on the leading edge of organic purchases, and coastal farmers had the same financial struggles as farmers anywhere. CROPP made a strategic decision to serve this burgeoning consumer market on the West Coast and organize local farmers into organic dairy pools to meet the demand.

The Cooperative had already marketed its organic cheese, butter and milk powder on the West Coast, so recruiting farmers like Doug Sinko to join a local pool to serve the region was the next key step.

CROPP's leaders embarked on this ambitious plan far from their Midwest home, because they believed the Cooperative could secure a sustainable pay price for organic farmers in this region, just as they had for their farmers in the Midwest. In this fast-growing market that was sure to become increasingly competitive, farmers without clout would probably be pressured to accept less for their milk.

In 1997, CROPP established its first West Coast organic dairy pool in Oregon and a second pool in California one year later. There were times in this regional strategy when markets developed faster than CROPP could organize pools. CROPP shipped milk to those regions where there wasn't a local farmer presence yet because consumers wanted organic milk. It was a temporary compromise, giving CROPP time to build its pools.

"We took milk from Westby, Wisconsin, to New York and then we shipped it across the country to the West Coast. It was absolutely insane, but that's how we got started." Louise Hemstead, CROPP chief operating officer

A Pioneer in Oregon

"I watched small dairies go by the wayside, and I realized we had to do something different," says Doug Sinko, today CROPP's Northwest Region dairy pool coordinator. "I earned my college degree in biological sciences and organic farming felt like a natural fit for us. We were already pasture-based." When an organic cheese plant in the area sought milk suppliers, Sinko decided to go for it. Neither Washington nor Oregon had state organic certification programs for dairy, so Miles McEvoy (named deputy administrator of the National Organic Program in 2009) steered Sinko to Oregon Tilth, a pioneering certification organization. They didn't have dairy standards either, but Sinko and his veterinarian proposed a list of requirements they thought would ensure 100 percent organic production. The careful folks at Oregon Tilth studied the list, approved it and Sinko transitioned his farm at Myrtle Point to organic in 90 days. Sinko's guidelines were so good that Oregon Tilth adopted them for future dairy certification.

Soon after the Sinkos were featured in the nationally-distributed *Hoard's Dairyman* magazine in June, 1997, Sharon heard from George Siemon asking if they

← What better place to raise healthy children than on an organic farm? A glimpse of Ross and Amanda Thurber's Lilac Ridge Farm in Brattleboro, Vermont.

might be interested in learning about CROPP's plan for creating regional dairy pools. Sinko asked other local dairymen to join him and the Myrtle Point contingent traveled from southwestern Oregon to Portland where they met with George Siemon and Theresa Marquez. "George talked about cows, shipping milk, animal diseases and livestock care. He was knowledgeable," Sinko says. "He was totally sincere, down-to-earth and committed to sustainable family farm agriculture."

It would take a few years before CROPP had enough milk orders to justify the 260-mile milk truck trip from Myrtle Point to Portland, but it happened in 1999.

> "Here in the West, if you're a conventional farmer and you're not growing by 10 to 15 percent a year, you're considered a failure."
> Tony Azevedo, Stevinson, California

Piling Cows on Cows

Tony T. Azevedo, the son of an immigrant farmer from the Azore Islands, wanted to replicate his dad's success in California's San Joaquin Valley. But by the 1990s when Tony and his wife, Carol, were running the farm, the economics had dramatically changed. "The average conventional dairyman has no control over his pay price," says Azevedo. "If he wants to increase his income, he has to milk more cows, so he just keeps piling more cows on top of more cows. When we recognized that, we knew we were at a crossroads."

What magic number of cows would produce a decent living? Azevedo was milking 150. He asked two neighbors with herds of 300 and 600 how they were doing. They weren't making ends meet, either, and they were considering doubling their herds. Where does it stop, Azevedo wondered? No one knew the magic, profitable number. "I came home despondent," says Azevedo. "I didn't want to be the generation that sold the farm." (By 2012, the average conventional California herd had grown to 1,500.)

Carol Azevedo handled the farm's books and she saw a steep drop in profits when they paid for chemical crop fertilizers. Let's forgo planting the crops, she suggested, and pasture our cows instead. Tony resisted at first. After all, his Future Farmers of America education advocated "tearing out all the fences, borrowing money for the biggest tractor you could afford and growing oats, corn and alfalfa." That worked for a while because Tony's father, Antonio, had kept his soil well-balanced. Over time, however, the natural nutrients in the soil were depleted and the Azevedos needed costly commercial fertilizers to try to make up for the loss. The result was steadily diminishing returns.

Tony and Carol Azevedo started pasturing their cows and stopped using chemicals around 1993. Within two years, they paid off their large chemical fertilizer

← Oregon organic dairy farmers Doug and Sharon Sinko; Tony Azevedo and his trusty cow-steed; Tony and his son, Adam, completed the transition to organic in 1996.

← Jersey cows from the Bansen herd in Monmouth, Oregon

Jon and Juli Bansen →

bill and reduced their veterinary charges because their cows were healthier. "It's not about doing something new," Tony says, "it's about trying to do it right." He even fed his cows dried kelp, his grandfather's practice in the Azore Islands. Sure enough, kelp's minerals helped clear up minor illnesses. Today, Azevedo says his cows live up to about nine years, compared to a two-or three-year lifespan for a conventional milk cow.

Around 1995, the Azevedos had an unplanned visit from Dave Pinkham, a Yale-educated, Vietnam veteran who drove a VW van and wore short pants and sandals. CROPP had hired Pinkham to help find dairy farmers willing to convert to organic production in California.

"He was pure hippie," Tony says, "the opposite of me. He educated us in organic and he became a close friend." The Azevedos completed their transition to organic in 1996 and they were among the three organic dairy families who filled CROPP's first California tanker load in 1998. Since then, Tony and Carol Azevedo have introduced more than 40 farmers to organic, including the members of CROPP's original California dairy pool.

Where once he was just a crazy farmer following a fad, now Azevedo, nicknamed "Tony the Tiger" for his unrelenting enthusiasm for organic, advises some conventional farmers. "Guys come and talk with me, especially when they're in trouble financially with their dairies," he says. "I've gained a level of respect. Even today, when I tell farmers they can run a dairy without antibiotics and pesticides, some will say it's an outright lie. But some of those farmers are now asking, 'How is this organic dairying done?' A lot of what I do is describe our practices and stand behind what we're doing."

> "I thought organic was a big gamble. It turned out it wasn't a big gamble at all."
> Jon Bansen, Monmouth, Oregon

What Are You Waiting For?

Jon Bansen was farming in Monmouth, Oregon, when he asked his organic farmer cousin, Dan, if he had any reservations about organic. His reply was succinct: "What are you waiting for, you idiot?"

After studying organic practices and talking with George Siemon in 1996, Jon and Juli Bansen transitioned their farm and got on the CROPP truck in 2000. A sustainable, strong pay price for their organic milk was the first attraction. But there were others. Bansen's grandfather had a Jersey herd in the lush Redwood forests of northern California's Humboldt County. His cows grazed on plentiful rye grass, and he used none of the modern chemicals that were supposed to boost production.

"Grandad was an organic farmer," says Bansen. "He didn't use pesticides, herbicides, commercial fertilizers, antibiotics or hormones. His example had a big

← Gage (left) and Guy Stueve clown around on their Oakland, California, farm; Gage at sunrise

influence on me." Like his grandfather, Bansen grazes his cows today. They are healthier because of it, he says, and they produce more milk. "As an organic farmer, you can take your foot off the throttle from time to time. You can do things just for the sake of making things biologically sound or aesthetically pleasing," says Bansen.

One year after the Bansens put up their Organic Valley sign at the end of their driveway, a stranger drove down the 100-yard entrance past electronic sensors and over a noisy cattle guard. "He got out of his car, stuck out his hand and said, 'I wanted to thank you for producing your milk,'" says Bansen. "That was the first time I realized that organic is really a different thing in consumers' minds. I knew it was different in my mind . . . and certainly in our cows' minds."

Not So Wild After All

Milk from Lloyd and Nancy Stueve's farm in Oakdale, California, was also part of CROPP's first truckload from its new regional dairy pool in that state. It was Stueve's father, a second generation farmer, who urged Lloyd to go organic. The elder Stueve had started his career as a milker in California, then a cottage cheesemaker, creamery manager and owner of a dairy. "We were already doing rotational grazing, so organic fit our practice," says Stueve. "Some of my farmer neighbors chuckled. They thought my plan was wild and out of the box."

When the California farmers first met with George Siemon to discuss a regional organic dairy pool, Stueve had the same reaction as his peers. "George had long hair and sandals and we were farmers in rubber boots. I thought, 'Let's just see what he has to say.'" The Stueves were certified in 1997 and joined CROPP the same year.

Nancy Stueve, a city girl from southern California, embraced the life. "Going organic really opened my eyes," she says. "I've learned what it means to be a good steward of the land we've been blessed with. It's not a nine to five job; it's a way of life and we've adapted well." Three generations of the Stueve family are now involved in running their organic dairy farm.

> "I was getting tired of the conventional world. I didn't know what my milk price would be. Every time I turned around, I was told I needed another chemical."
> John Boere, Modesto, California

"I Can Sleep Well"

Wayne Peters, Tim Griffin and Tony Azevedo drove the back roads of Northern California near Modesto looking for more dairy farms with pasture-fed cows. These farms were the most likely candidates for organic dairy production.

Peters, Griffin and Azevedo slowed down when they saw the upside down co-op sign. The positioning was no accident. John Boere and his family had a 200-acre farm with about 500 milk cows in Oakdale. When his smaller cooperative, California Gold, sunk into a financial quagmire and sold to a much larger co-op, Boere was disillusioned. "The new owners were big. They were all over the country. Running the co-op was 'their way or the highway.' The members had little or no influence," says Boere.

He upended his co-op sign in protest. Though it would take time and expense, Boere was a ready convert to organic dairying. He had heard about a dairy farmer losing more than 30 cows when the chemically-treated alfalfa bales he bought killed his cows and made other cows sick. A neighbor had been imprisoned for misuse of antibiotics with his beef cattle. "Being organic means that my family is safer, my cows are safer and the environment is safer," he says, "and I can sleep well at night not worrying about my milk having any antibiotics or chemicals in it." With a stable milk price that he can count on, Boere remains a farmer. "I would probably be out of business by now," he says, "if I hadn't converted to organic dairy farming in 1999."

In 2003, on average, American prepared meals contained ingredients from at least five countries.

Here a Pool, There a Pool

CROPP's geographic expansion outside the Midwest was not solely on the West Coast during the second half of the 1990s. A group of Maine organic farmers became CROPP's first eastern dairy pool in 1997, and Vermont and Pennsylvania followed just two years later. By the end of 2000, CROPP would have organic dairy pool members in 12 states. With all that new organic milk production, CROPP needed a tireless, impassioned professional to lead national sales. If CROPP was going to give more organic farmers a fair pay price for their efforts, the Cooperative had to successfully sell its organic products in markets all over the country. Theresa Marquez recruited Eric Newman to be the Cooperative's first national sales manager. A native Californian and organic advocate, he was living with his family in Bozeman, Montana.

Newman owned two restaurants in Montana and co-managed the Community Food Co-op in Bozeman, where he was instrumental in moving the Co-op to Main Street in that thriving town. Newman had been interested in nutrition and organic since his youth, but having two young children of his own convinced him that organic agriculture was essential for good health. Newman became President of the OCIA (Organic Crop Improvement Association) Montana Chapter #1 and, through this work, met a broad spectrum of organic farmers in Montana and the United States.

"I connected to the farm side of food production and began to really understand how critical livestock production is to organic agriculture," says Newman. When he met Siemon and Marquez at the Natural Foods Expo in Anaheim, California, Newman didn't know Siemon's role at CROPP. "He was this unassuming farmer. A humble guy who didn't say anything about being CEO." Newman was already selling the Organic Valley brand and was committed to Cooperative principles.

"CROPP was willing to pay about $32,000 for a national sales manager, the same amount I was making in Bozeman," says Newman. "I think I was one of a few people willing to take that wage and move to rural Wisconsin. I saw it as a terrific opportunity. I sold newspapers as a kid, yearbook ads in high school and learned to sell in the restaurant business. I didn't have 'traditional' national sales credentials, but I was willing to work hard and understood the retail natural foods business."

Newman and his family started their trek from Bozeman to Viroqua, Wisconsin, in February, 1996. Towing a Volvo with their Isuzu Trooper, Newman, his wife, Anna, and their two children, Will and Nico, left Bozeman in a -30°F deep freeze. As they crossed into Wyoming, the Trooper lost all of its transmission fluid and the nearest garage mechanic told Newman to drive on. "It'll be weeks before I can fix your car," the mechanic said. "Just put the car in fourth gear and keep driving. It doesn't really need any fluid to get you there."

They navigated the curving, barely-lit, narrow and ice-covered back roads near Chaseburg, Wisconsin. At last, in brutal cold, the family arrived at Robert and Margaret Siemon's farm house, next door to George and Jane Siemon. Inside, the pipes were frozen, the firewood was wet and the Siemons were busy making the house habitable.

"That was our welcome to CROPP," Newman says, "and I never looked back. Mike Bedessem had a list of farmers on a reserve pool list and he said, 'Eric, look at this list of 35 names. We're waiting to bring these farmers on. You've got to sell more milk.'" That's just what Eric did, by assembling a small but tireless team to do it. Newman set the tone by being the most tireless of all.

The Stonyfield Connection

CROPP's entry into New England was linked with the growth of Stonyfield Farm of Londonderry, New Hampshire.

The organic yogurt company had its start in 1983 with a recipe developed by Samuel Kaymen, a rural development visionary. By 1997, Gary Hirshberg, Stonyfield's "idea man and CE-YO," and his sister, Nancy, a champion of organic, had built Stonyfield

↑ CROPP farmer-members in California (from left): Clarissa Coelho, John Boere, Elainna Coelho, Frank Coelho, Tony Azevedo, Gage Stueve, Chantelle Coelho and Frank A. Coelho; John Boere on his California dairy; in the CROPP kitchen (from right to left): Eric Newman, Jerry McGeorge, LeSoung Bateman and Zell Spry; Eric and his son, Nico, on a tour of Richard deWilde's Harmony Valley farm in 1997

to a multi-million dollar enterprise. Nancy Hirshberg had pressed for Stonyfield's conversion to all-organic yogurt, and CROPP had begun supplying them with organic milk powder.

"CROPP joined forces with Stonyfield and we both grew up together." Eric Newman, CROPP vice president, sales

By the late 1990s, Stonyfield was looking for more regional organic milk suppliers. The Hirshbergs asked CROPP to organize state pools in New England and, over time, CROPP would become Stonyfield's sole supplier of organic milk (in fluid and powder form), representing more than 25 percent of the Cooperative's total milk sales.

Organic milk from Gregg and Gloria Varney's dairy farm in Turner, Maine, was in that first truckload of fluid milk bound for Stonyfield Farm. The Varneys had become Maine's first certified organic dairy farmers in 1993, and they joined CROPP in 1997 as members of the Cooperative's first dairy pool in Maine.

Varney is a third generation farmer who managed the 120-cow family farm after he graduated from college. He gave up his father's practice of using chemical sprays in 1976. It wasn't easy being an organic farmer in his region. "Up until recently, farmers judged other farmers by how clean and free of weeds their acreage was," says Varney. "If you had weeds (which is common with organic farming), you weren't really progressive. We were viewed as hippie farmers living a fad that would surely pass."

"Can we feed the world organically? Of course we can. We always have." Gregg Varney, Turner, Maine

Varney says he was impressed that "CROPP's Board of Directors based in Wisconsin recognized that it costs more to produce organic milk the farther east you travel." CROPP backed that up by offering Varney more for his milk than CROPP's midwestern Board members paid themselves. The farther a state is located from a dairy industry center, the more costly it can be for a dairy farmer. Feed and equipment prices, for example, tend to be higher. "They were extremely fair with us from the beginning," he says.

Varney likes the idea that CROPP held its milk pay price firm, even in the face of downward price pressure from competition in the organic marketplace. He attributes CROPP's price-keeping success to controlling its overall milk supply from farmer-members and not buying more milk than it can sell.

← Nancy Hirshberg of Stonyfield Farm in 2000

Gregg and Gloria Varney in Turner, Maine, and their son, Roy, gathering eggs →

Steve and Julia Russell in Winslow, Maine →

"The stable pay price helps farmers deal with all the other variables in their lives. Things happen. That's farming."
Travis Forgues, Alburg, Vermont dairy farmer and CROPP Board member

"What's Unique About Us"

Travis Forgues didn't even plan to follow his father into farming. When Forgues graduated from college in the early 1990s with a degree in computer science and psychology, the agricultural economy looked bleak. But Forgues' father had introduced intensive grazing to the family's dairy herd and, in the mid-1990s, father and son began exploring other ag alternatives, including organic farming.

Organic Cow of Vermont, a small independent business, was recruiting new farmer-members, and so was CROPP. Forgues petitioned his father to go organic, but one day while he was away, "soft-spoken, amiable Jim Wedeberg of CROPP, a guy any farmer could relate to," won his father and mother over to organic in an afternoon visit to their Vermont farm home. "I came home and my parents were talking about building a load of organic milk for CROPP," says Forgues. "Jim's presence just changed their thinking about organic."

Getting on the CROPP truck wasn't easy, even though the Forgues farm quickly qualified for organic certification. As it turned out, CROPP's organic farmers in Maine had mobilized a little faster than Vermont's farmers, so CROPP had all the milk it needed to supply Stonyfield in New Hampshire. In his put-the-farmers-first style, Jim Wedeberg counseled the Forgues family to supply Organic Cow of Vermont, because CROPP couldn't guarantee how soon it could take on new members.

That was the delicate balance that CROPP faced in growing organic markets around the country and creating dairy pools to supply them. If supply preceded market demand, some farmers had to wait to join CROPP. If demand exceeded local supply, CROPP had to step up its farmer recruiting.

Organic Cow was the first organic milk in New England and New York metro supermarkets, and had engendered lots of loyalty. Before long, however, Organic Cow sold to Horizon Organic Dairy of Colorado, the fast-growing milk company that was CROPP's customer *and* a market rival. "We felt like Organic Cow was ours, but it really wasn't," Forgues says. "At the end of the day, one guy had the power to sell us and get a check. Horizon was large and I didn't want to be part of that."

Instead, Forgues and two other organic dairy families joined together in 1999 to produce 10,000 pounds of milk every other day for CROPP with the promise of more to come. By this time, the organic yogurt market had grown significantly, and Stonyfield needed more

Vermont farmer-member Travis Forgues with his children, Gabriel and Molly; Travis with his father, Henry →

milk. "I agreed to help recruit farmers," says Forgues. "We grew enough to supply our second load of milk also in that first year."

Now a member of CROPP's Board of Directors, Travis Forgues represents an organization that is itself approaching $1 billion in annual sales. "We've gotten large, that's true," says Forgues, "but CROPP's farmers are the owners, and we stay connected to them. We treat them with respect, and we have ready access to anyone in the organization. That's what's unique about us."

Regina Beidler and her husband, Brent, would second that statement. The young couple had been married three years when they learned about dairying in their home state of Vermont and became one of the first farms from their state to join CROPP in 2000. The Beidlers had plenty of coaching from friends and fellow farmers Bob and Beth Kennett as the young couple learned to raise 15 heifers, the start of their dairy herd.

"Going organic was an obvious step for us when we began to develop our farm plan," says Regina. "Sustainability is a word we all hear often. For us, it's a word that begs us to determine how to live and farm in a way that will support not only the short term, but also generations to come."

Regina and Brent Beidler had a chance to learn more about sustainability when they volunteered through their Mennonite church to establish a small scale agricultural demonstration project in Chad, West Africa. In the small village of Bitkine, they introduced young people to raising onions and garlic, while providing seed for other farmers. "In the rural villages in a country so vast, no police forces or government agencies exist," she says. "People rely on tribal structures that have been in place for centuries. There is a realization that the good of the whole is more important than the good of the individual."

The Beidlers were attracted to this same philosophy of cooperation at CROPP. Regina, Brent and Erin, their daughter, share daily farm chores with diligence and joy, even when the work is hard. "We understand that our relationship with the land is as sojourners that have ownership for only a short time," Regina says. "Stewardship and sustainability are at the forefront of each decision we make as parents, as farmers and as community members."

More Than Milk

In the late 1990s, along with organic dairy farmers, CROPP also welcomed more farmers to its vegetable, egg and juice pools, as well as its fledgling organic meat pool (this was CROPP's second run at producing and marketing organic meat). With farmer-members producing increasing numbers of organic products, consumers around the country had more plentiful choices. It wouldn't be long before the CROPP acronym would reflect the Cooperative's expanding geographic reach.

← Vermont farmer-member Brent Beidler with his daughter, Erin

CROPP added more farmers to its juice and egg pools in the late 1990s. →

How Hard Can It Be?

The Amish and Mennonite communities in Northeastern Ohio did some soul searching in the late 1990s. Though the area had long been a haven for small dairy farms, they saw their numbers depleting as some of their children lost interest in farming and good land became increasingly scarce.

"We met in our farm shop in 1997 to discuss how we could add value to our farm products and romance our young people back to farming," says David Kline, Amish leader, author and publisher of *Farming Magazine*. "We decided we needed three things: One, we had to make money because conventional farming was a losing proposition. Two, we couldn't be overwhelmed by farm work all the time. And, three, farming had to be enjoyable again." The group of about 11 farmers decided to go organic, even though Ohio had no organic dairy standards in 1997 and CROPP didn't have the customer demand in that state to take them on.

OK, the farmers decided, let's start our own co-op. How hard can it be?

The ringleaders were brothers, Mark and Ernest Martin.

The eleven Amish and Mennonite families that formed Ohio Family Farms Association called their brand Family Fresh and they were the first organic dairy producers in the state. They arranged to have their milk bottled and sold to Auburn University. Their milk also appeared on the shelves of popular Mustard Seed stores around Ohio. But when they called supermarkets in major cities including Columbus, Cleveland, Toledo, Akron and Canton, the response was disheartening: "What *is* organic?"

"Only the natural food stores understood," says Mark Martin. When he explained organic, the answer usually was, "Sorry, we don't have a market for that."

Jim Wedeberg at CROPP was an informal coach for the Martin brothers. "I called Jim regularly if we had a question or problem," says Mark. And there were plenty of both.

The fledgling cooperative couldn't find anyone to pick up and haul their milk, so they bought their own milk truck. Then they were plagued with bacteria in their milk because the driver forgot to wash the pump hose after each milk pick-up.

Truck and driver mishaps made their work even harder. On Christmas night, their truck driver was headed to Utica, Ohio. He called to tell the Martins their truck—with milk on board—had broken down. Mark found a substitute hauler at 11 p.m. "It was extremely stressful," he says, "and we were desperate."

Meanwhile, back at headquarters, the Martin brothers handled all the co-op's paperwork by hand, computing pounds of milk produced by each farmer, tracking butterfat and protein levels, keeping books and writing checks to farmer-members.

The co-op also sold Family Fresh organic cheese, but they couldn't sell it fast enough before it became overripe. "We gave some away," says

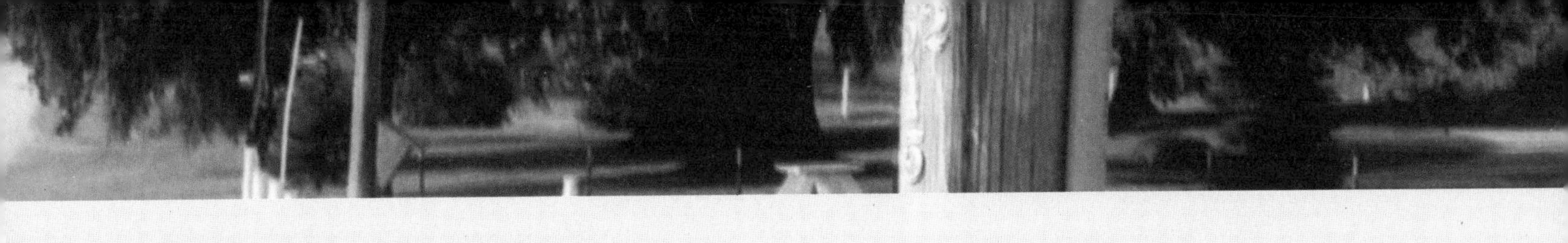

Ernest, "and sometimes we just threw it away."

When a co-op member left, overworked and burned out, others had to take on unfamiliar tasks. Making labels for the organic milk bottles was one. Ernest bought a batch of labels in the right size, but didn't have a printer to make them. He searched Mansfield, Ohio, settling for a coin-operated printer in a camera store. "I wasn't quite sure how to operate it and I ended up breaking a very expensive machine," Ernest says. He left contrite—and empty handed—and he later settled for simpler, but unattractive, black and white labels from an ink jet printer. The little co-op also got into a tussle with the State of Ohio about what kind of language they could put on their labels. They were not allowed to use the terms "no GMOs" or "no hormones."

There was considerable turnover in the co-op's tiny paid staff, too, especially in sales and distribution. The operational headaches morphed into migraines.

Though farmer-members were passionate about the project, the mistakes, stresses and overwork were hard on their families. "We could offer a course in how *not* to start and run a cooperative," says Ernest.

The Martins asked Jim Wedeberg at CROPP again, "Is there any way CROPP can take us on now?" The answer was "No, not yet; hang in there."

Finally, within a few weeks of each other, Horizon Organic called the Martins asking if Horizon could buy their milk *and* Jim Wedeberg proposed a conference call with CROPP to discuss possible membership.

The evening conference call was in Ernest's home because he was the only one with a speaker phone. Nervously, he dialed the number and connected to La Farge, Wisconsin. Just as the conference was about to begin with Wedeberg and George Siemon, Ernest's young son, Randall, tripped over the phone cord and disconnected the call. Panic. Redial.

The call ended with a plan for Jim and George to visit with the co-op's members in Ohio. Though not all of them were initially in favor of joining CROPP (some preferred independence), the group ultimately agreed that all or none should become members. The 11 farms had an average of about 40 cows and together they produced about 30,000 pounds of milk every other day, enough for a full truckload.

↑ Organic Valley staffers toured the farms of Ernest and Mark Martin in Shiloh, Ohio.

"I cheered that first morning the Organic Valley truck came into our driveway," says Ernest. "We had put so much focus on our co-op that we were neglecting our own farm work. We got less than the conventional milk price, just to get the co-op off the ground. Joining CROPP kept me in dairy farming."

Today, Ohio has the second-most dairy farmers in the Co-op, after Wisconsin.

CROPP's new name, introduced in 2001, stood for Cooperative Regions of Organic Producer Pools. CROPP retained the same five letters from its founding name, but it now had a dramatically different "organic footprint" that stretched to both coasts, and even beyond that.

"CROPP is now a seasoned warrior in the complex task of balancing organic supply." George Siemon

A Stable, 3-Legged Stool

With all of this geographic expansion in the second half of the 1990s, CROPP's annual sales spiked from $13.8 million in 1996 to $72.6 million in 2000. To a large extent, this growth reflected a foundational marketing strategy that CROPP had adopted at its inception. They call it "the 3-legged stool" (think milking stool). At the heart of this strategy is product diversity.

Leg one of the stool includes CROPP's branded sales and private label—the largest, single revenue producer.

Leg two is ingredient sales: specially-designed ingredients for manufacturers (tons of dry milk powder for Stonyfield Farm and Organic Valley cheese for Annie's Homegrown Mac & Cheese, for example). These organic ingredients are often the by-products of branded production.

Leg three includes raw, bulk organic sales: whole tankers of milk, pallets of squash, dozens and dozens of eggs fresh off the farm that others will use in food processing. By achieving this diversity of product sales, CROPP is better able to balance its members' perishable production with market demand. Not-so-quickly-perishable products like butter, cheese and dry milk are essential in the balancing act that CROPP faces daily. Three legs of the "product stool" make for a stable balance of supply and demand, but remove one leg, and the stool tumbles.

There is also a fourth leg of the stool, though sales in this category are very small. It is the sale of excess organic production on the conventional market. This outlet allows CROPP to sell the intermittent tail-end of its production (though at a lower price), thus achieving near 100 percent utilization of its production. That's called "balancing," and it has become a fine, disciplined science at CROPP.

According to U.S. Census data, fewer than 30% of businesses see their 10th anniversary.

There was a lot of balancing to do in the second half of the 1990s. During that five-year period, the Cooperative's farmer-membership had nearly tripled, from 119 to 340, and so had CROPP's staff, which had grown from 50 employees in 1996 to 148 in 2000. This meant that by the end of the decade, CROPP was supporting the livelihoods of nearly 500 families.

"CROPP is becoming increasingly complex. This presents more challenges to the goal of delivering the vision we all hold true to. It is crucial that the farmer-owners . . . stay informed and involved." George Siemon in 1998

No Pain, No Gain

When CROPP reached its tenth anniversary in 1998, George Siemon was frank. "Any business that grows $8 million in a year can count on growing pains. If 'no pain, no gain' is true, then we can say our gains were well earned!" All that fast growth made access to capital a constant worry and the size of the enterprise called for careful oversight. "As CROPP grows, it's crucial that we all understand the structure . . . and how these pieces of the puzzle work to keep CROPP a true farmers' cooperative," Siemon said. While "bigger" meant more authentic organic choices for consumers and more farmers earning a fair pay price for their production, becoming one of the "bigs" in this fast-growing food sector could turn them into the kind of organization they abhorred: bottom-line-driven, unresponsive, careless with people and products, greedy.

In that anniversary year, farmer-members earned a record average pay price for their milk—$17.92 per hundredweight—and a record $1.25 a dozen for egg pool members. In fact, the egg pool doubled its production that year. The largest number of new vegetable growers (the most in four years) also joined CROPP in 1998. When CROPP successfully mobilized organic farmers to lobby the U.S. Secretary of Agriculture, the government finally blessed organic meat labeling in 1999. That gave CROPP's young meat business a boost. No wonder the State of Wisconsin named CROPP the state's top rural development initiative.

Ultra Pasteurized: A Game Changer

The pace was exhilarating and exhausting for Eric Newman, CROPP's Vice President in charge of national sales. Newman and his staff had organic dairy, vegetables, eggs and meat to sell to retailers, food companies and institutional buyers, plus a blockbuster new product: Ultra (high temperature) Pasteurized (UP) milk, introduced by CROPP in 1998.

It was the opportunity that opened up the U.S. for CROPP's organic milk, and it was their entry into big grocery chains. Up until 1998, CROPP was limited in how far it could ship its traditionally pasteurized organic milk because of the product's limited shelf life.

> "When I told Albertson's that we could deliver ultra pasteurized organic milk to all of their warehouse locations with 30 days' minimum code date, they liked that idea a lot." Eric Newman

In the Pacific Northwest and the Midwest, CROPP had local dairies to process and bottle their milk for distribution largely to stores in the region, but what about other regions like the heavily-populated East Coast? People knew the Organic Valley brand in the East because they bought CROPP's organic butter and cheese. What about milk?

Enter Stonyfield Farm's Nancy Hirshberg, a tireless proponent of organic dairy and family farms, who would become one of CROPP's greatest friends and partners.

← Organic Prairie package shows the USDA Organic seal.

CROPP introduced Ultra Pasteurized (UP) milk in 1998. →

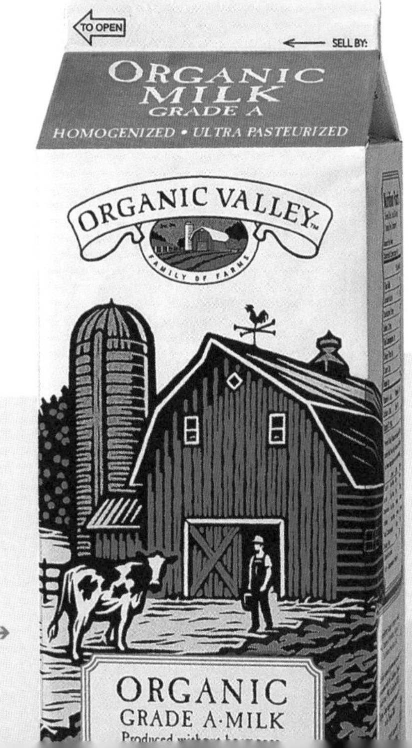

"Nancy had been talking to the people at Wakefern cooperative in New Jersey [ShopRite banner stores] about milk spoilage. She came to us with an idea," says Newman. CROPP could produce its new Ultra Pasteurized milk at the "Ultra Dairy" owned by Morningstar in New York, which was outfitted for this special pasteurization process. Then CROPP could distribute the milk up and down the East Coast, starting with Wakefern. No one else in the organic business was doing it—yet.

CROPP was the first organic dairy to adopt this new method of pasteurizing milk. The milk is initially preheated, then—within a second—steam is injected into the milk at 280°F for two seconds. Finally, it is flash-chilled. This process destroys virtually all bacteria in the fluid milk and preserves its freshness and taste. This milk has a shelf life of up to 70 days—three times longer than ordinary pasteurized milk. (Traditional pasteurization involves heating milk to 165°F for 16 seconds leading to a shelf life of 21 days.)

Selling to Wakefern and then supplying other large grocery chains and natural foods stores around the country was a dream come true for CROPP (meanwhile, their biggest rival, Horizon, was publicly skeptical of UP milk).

Typically cautious, George Siemon and Mike Bedessem were concerned that the Co-op might not be able to sell all the UP milk that Ultra Dairy processed on the East Coast. The solution they came up with was to partner with Stonyfield Yogurt. They could share the costs, sales efforts and success. Stonyfield would focus on the New York metro area north and CROPP would pursue everything south of New York. When Stonyfield passed on the opportunity, CROPP swallowed hard and went ahead on its own.

By the spring of 1998, CROPP was shipping close to 1,000 cases of organic UP milk from the New York plant to natural foods stores, largely on the East Coast. Newman, joined by John Morrisey in sales, made presentations to virtually every major supermarket chain on the coast representing a combined total of about 7,500 stores. Newman next pursued all the other major chains in the U.S. The orders came rolling in and by 2000, CROPP was shipping more than 60,000 cases of UP milk all over the U.S., including Alaska and Hawaii. They were delivering UP milk to all of United Natural Foods' warehouses across the country.

This blockbuster product represented more than 50 percent of Organic Valley branded product sales. And at the end of 1999, when CROPP was reporting astonishing 66 percent growth over the previous year, the Co-op's UP milk category had grown by a mind-blowing 766 percent.

"Extended shelf life milk was a brilliant move for the time, even though it was counter-intuitive for the organic and natural food industry." Louise Hemstead

"Consumers said, 'Ultra Pasteurized? That means fresher, cleaner, safer and we love the shelf life.'" Eric Newman

A Panoply of Partners

All of this growth called for more partners who could process CROPP's organic production. By the end of 1999, Darigold and Land O'Lakes joined that partner list as processors of CROPP's UP milk. In addition, CROPP established new processing relationships with four cheese plants, one glass bottler for "cream on top" milk, four meat facilities, a yogurt plant, two butter plants, one milk-drying facility and an orange juice bottler. In all, CROPP had 47 trusted, certified organic facilities around the country working with them.

The Cooperative had sought out these partnerships from its earliest days when CROPP started with a dream but no resources. The founders built the business by developing a "healthy co-dependency" on the resources and skills of others, and this led to many long-term, mutually-beneficial partnerships. "Our philosophy was always to build the business and then the buildings," says George Siemon. "We built CROPP by identifying co-pack plants to work with us. We rented warehouses and offices until an obvious need was well-established. Only then would we consider building our own facility."

During the 2000s, Upper Midwest dairy farms discontinued production at the rate of more than three per day.

Put Your Money Down

In 1998, nine years after buying the abandoned cheese plant in La Farge, Wisconsin, and turning it into their humble headquarters, CROPP bought its second building—a small, shuttered creamery in Chaseburg, Wisconsin, population 350. CROPP would convert into a butter plant the little creamery that had served as a destination and gathering place for dairy farmers and milk haulers for six decades. By doing so, CROPP could boost its butter production and exert more quality control over the butter-making process.

The indomitable Wayne Peters, CROPP's "just-do-it" dairy pool co-founder, made the purchase possible. CROPP didn't have the $750,000 purchase and renovation price for Chaseburg sitting in petty cash. The financing journey started when a woman, whose husband was a milk hauler, asked Wayne partly in jest, "How can I get you to buy the Chaseburg creamery?" Wayne's answer was practical. "Talk to your boss at the bank and see if he can give us a good interest rate." She did.

The representative of River Bank of Stoddard, just west of Chaseburg, asked Peters for a tour of the creamery and a description of what CROPP planned. Peters had a soft spot for Chaseburg, located just down the road from his own farm. The vacant building on Main street was a worry for the townspeople. After the tour, the banker was blunt. "I don't know anything about Organic Valley," he told Peters. "I don't even know if there *is* an Organic Valley. The only one I know is you."

Peters did some fast thinking and invited the banker to La Farge to meet CROPP's management team. The banker met the team and seemed favorably disposed to a loan: here was an organization willing to invest in rural communities.

On the trip back to Stoddard, the banker and Peters remembered they knew each other from the days when Peters sold farm equipment for Rochester Silo. They also discovered that one of Peters' relatives was related

← Wayne Peters and his beloved Chaseburg Creamery at its grand opening in 1999

Wayne and Brent Roiland inside the new Chaseburg Creamery butter plant that same year →

"I Was Enamored"

Virtually every *CROPP Report* newsletter in the late 1990s had a "new faces" section as the Cooperative employee pool grew. Someone needed to take the lead in human resources and that person was Jerry McGeorge, the son of CROPP's first chief financial officer, Jerome McGeorge.

The younger McGeorge had no plans to join CROPP, though he had learned about organic farming as an eight-year-old. His parents bought a farm in Tennessee with two other couples and experimented with communal living and growing their own food without chemicals. "We lived in a house with no running water and no plumbing," he says. "We were doing organic farming before people had a name for it. I was enamored with the life."

McGeorge graduated from college with a degree in social work and moved to Chicago with his fiancé. He cared for babies born with HIV and crack addictions, and he worked in a school on the south side of Chicago with poor, struggling families. When Jerry and his wife, Susan, started their own family, they left the big city and moved to Vernon County, Wisconsin, where Jerry intended to pursue law school. He had no intention of joining CROPP until the persuasive Theresa Marquez urged him to sit down with George Siemon in 1997.

Siemon's words riveted McGeorge. "I know your background is social work, Jerry," Siemon said, "and you've told me how important that is to you. I don't think there's any clearer social work than the work that CROPP is doing."

McGeorge agreed. "CROPP was producing the healthiest, most nutritious food for people and telling them why it was an important choice to make in life," he says. "They set out to create a successful alternative to conventional agriculture. They were committed to keeping families on their farms and to maintaining a way of life that went back generations."

The young father thought back to the youngsters he worked with in the poorest neighborhoods of Chicago. If their moms packed lunches for them, the choices were often chips and sugar-laden snacks. He wondered if their problems with learning and concentration were affected by the food they ate.

But what about law school? "Take a year; work with me," Siemon told McGeorge. "If you think you still want to go to law school after that, it'll be waiting for you."

"When I started, my desk was two-by-fours with a plywood sheet nailed on top. When I hired my first person, we added four more feet of plywood to my desk for her."
Jerry McGeorge

With no business experience, McGeorge had a crash course in co-op policies, financial priorities, supply management and farmer and employee relations. It was up to him to help create a positive working environment at CROPP that reflected the core values of the Cooperative's founders.

by marriage to a key River Bank staffer. The banker recommended approval of the $750,000 loan on one condition: that Wayne Peters supervise the renovation and write all the checks. "That kind of ticked off George, Louise and Mike," Peters chuckles, "but sometimes it's not about what you know, but who you know."

Within two weeks of starting operation, the Chaseburg butter plant exceeded the original weekly plan to produce 24,000 pounds of butter and 13 truckloads of milk. And because of CROPP's quality assurance, the Co-op reduced its product loss costs by $145,000 in the first six months. In 2012, Chaseburg's weekly average was 150,000 pounds of butter and 400 truckloads of milk.

> "CROPP can be humble and hardworking as well as prosperous and magnetic." Wayne Peters, CROPP 1999 Annual Report

We're Not for Sale

Like a good scout, CROPP was humble and hardworking. No one in the organic industry could overlook the success story of the "Little Cooperative That Could." Managing a fast-growing organization is the most challenging of all assignments in business and cash shortages can kill. Though CROPP's sales exploded through the late 1990s, the cost of doing business was high and 1997 was a losing year. Mike Bedessem, CROPP's CFO, remembers. "For all the phenomenal growth, it's hard to run a business that grows that fast while still remaining independent." Horizon and other companies were interested in buying part or all of CROPP. "It seemed that every other year we were receiving an offer and we always said, 'No thanks.'"

But one of those years in the late 1990s, CROPP's leadership *did* consider an offer from Horizon. On the plus side, if CROPP sold, it would be owned by a larger entity with deeper pockets and more access to capital. The farmer-members, investors and management team would earn a handsome return and the farmers would also be assured of a long-term contract for their production.

As clearly documented in the financial pages of newspapers from that era, mergers and acquisitions seemed almost epidemic. Few apologized for selling out. It was the smart thing to do. If CROPP remained independent, however, it could continue building the Cooperative in its own way for the next generation of farmers, keeping its ideals and core values intact.

After days of discussion, CROPP's management team went around the table considering Horizon's offer. One by one, each person spoke. "I'm not interested," was repeated over and over, says Bedessem. CROPP's Board of Directors backed the management team's recommendation and talk of selling was never seriously discussed again.

> "Our Board decided they would not sell away the rights of the next generation. That was the big decision." Mike Bedessem

CROPP's leadership would still talk to would-be suitors, however. CROPP was gracious *and* inquisitive in those meetings. They took the opportunity to gain as much insight into their industry and their competition as possible. How could CROPP increase sales? How might they bolster their bottom line? It was a chance for a closer look at their competition's turf. As George Siemon said, "It's free to ask questions."

Free At Last

"I'm glad that we stuck to our principles," says Eric Newman of the decision to remain independent. "We were naïve, or brave enough, to think that we could do this on our own."

← A meeting of the CROPP management team in 1999 (from left): Louise Hemstead, Cecil Wright, George Siemon, Mike Bedessem and Eric Newman

Employees with new children demonstrate that farm families weren't the only families that CROPP provided for as it grew. →

Yes, but not without help and considerable creativity.

In the late 1990s, Mike Bedessem faced CROPP's financial problems at close range. Cash flow was so tight that one of CROPP's major bottling partners called and said it would not handle CROPP's milk the next week unless they were paid. Those were dark days. While he never mixed his business career with his faith, things had gotten so bad, Bedessem turned to prayer.

"I had a rosary that my family brought back from Yugoslavia, and I decided to say the rosary while I drove from my home in Gays Mills to La Farge, rather than listen to public radio and drive myself crazy with worry," Bedessem says. "I was praying, squeezing those beads as hard as I could. I prayed that enough money would come in that day because CROPP was in real bad shape."

Bedessem arrived at CROPP headquarters, to discover that they had just received the single, largest deposit in the Co-op's history: $1.2 million. CROPP paid its bottling partner pronto.

CROPP's relationship with its banker had become tenuous, too. "We were growing so fast and our equity to asset ratio had fallen to 16 percent," says Bedessem. "By comparison, our ratio today is about 60 percent." CROPP's bank, the St. Paul Bank for Cooperatives was not willing to renew CROPP's operating loan unless the Cooperative agreed to sell part of its fledgling meat business. CROPP would not. That's when Bill Bosshard, CROPP's banker from the early years, offered bridge financing to tide them over.

During this troubling time for CROPP, Mike Bedessem stood in the shower one morning thinking, "Bill Bosshard is willing to carry us, but he insists that we find financing to replace his bank. We've gone to other banks and they turned us down. We've got to get freedom from these banks." That same morning, Bedessem called Ron McFall, CROPP's attorney. McFall described a trust that investors could put money into, earn a high return (12 to 14 percent) and the trust would then loan the Co-op money.

Bedessem called it "the Freedom Fund" and the response was, well, freeing. About 50 people invested. Outside investors earned a hefty 14 percent on their money, and farmers and employees earned 12 percent. "Some people just invested because they believed in organic," says Bedessem. "I remember getting a call from the CEO of a New York Stock Exchange-traded company. He sent us $200,000 the next day. A CROPP farmer-member invested $500,000. Other CROPP farmers invested and so did employees. Even Bedessem's father invested.

Within three months, the Freedom Fund raised $2.8 million. "We continued the program for about three years," says Bedessem, "and it gave us the working capital we needed to continue to fund our growth."

Now, that's freedom.

Farm Women * *Regina Beidler*

I have long been an admirer of farm women. Though the number of women taking the lead on farms has doubled since the 1970s, women farmers often work quietly beside their partners, milking cows, feeding calves, driving tractors, cleaning barns, chasing escaped heifers and unloading hay as well as managing households and pinch hitting wherever needed.

I came to be a farm woman by marrying a man born to be a farmer. Although each of us finds our own unique path based on our abilities and interests, I was also fortunate to have strong farm women in my family to look to for inspiration.

Both my mother and mother-in-law grew up on farms in eastern Pennsylvania. My grandparents raised beef, potatoes and tobacco for sale, and had a flock of hens for small-scale egg production. They had several cows, which they hand milked, as well as other small livestock and a garden that provided for the family's needs. I asked my mother what specific memories she has of the roles that she, her sisters and mother took on the farm. She is quick to say that everyone, men and women, assisted with all the tasks on the farm. Children were initiated into the work by being responsible for bringing feed and water to the "peepy house" where the young chicks began their lives, and to the pasture shelters where the hens spent their days.

By the time my mother was ten years old, she was responsible for milking the cows, washing and grading eggs, picking up potatoes and assisting with unloading hay into the barn. As a teenager, my mother transplanted young tobacco plants with her siblings and drove tractor when needed.

My husband's grandmother, with help from her husband, raised thirteen children on a 125-acre dairy farm. Over her lifetime, Grandma committed hundreds of poems, stories and scripture passages to memory, which she would recite to the delight of her children, grandchildren and great grandchildren. She also loved the farm and not only worked side by side with her husband but was the inspiration that helped make their farm a success. She loved dairy cows. One can imagine the relief it was to her to escape to the barn for milking and a little time alone with her own thoughts.

The farm had a diverse animal population, including cows, chickens, geese and sheep. Every year Grandma raised 100 turkeys which she sold during the holidays in order to buy Christmas presents.

In his review of the book, Vermont Farm Women, *Jeffrey Lent comments, "Implicitly, for generations farming women have enjoyed true equality, not only in work but also in the fortune and beauty as well as the grief and uncertainties of this life."*

For those of us who live this life, that is truth.

{ Vermont Organic Valley dairy farmer Regina Beidler, along with her husband, Brent, and daughter, Erin, daily milks a pampered herd of organic cows in their 200-year-old barn. Beidler has been a CROPP dairy farmer and active organic advocate for 11 years. }

Partnering the Organic Way

* *Nancy Hirshberg*

Glover, Barton, Bethel, Troy. As we passed through town after town in the dark hours of early morning, despite our lack of sleep, George Siemon and I felt energized. We hadn't expected such a turnout—or such interest—from the group of thirty or so dairy farmers who, hours earlier, had come to the local Grange to learn about how to go organic.

"What's to prevent you from pulling out and leaving us high and dry once we've invested in going organic?" "What do I do if my cow needs an antibiotic?" "How do we know you won't lower the pay price once I go organic?" Despite the emotion in their voices and their prodding questions, their optimism—while cautious—was palpable.

It was 1997. Dairy farmers had been through some lean years and almost always got the short end of the stick. Failed government programs to manage supply and farmer pay price, farm cooperatives forgetting that they represented the farmer-members, and decades of "get big or get out" mentality in agriculture meant the small dairy farmer was being squeezed out. For many, the joy of farming was fast disappearing.

We had been making yogurt in New Hampshire for fifteen years, first on our organic farm and then when we outgrew the farm, in our Yogurt Works. No longer dairy farmers, we were seeking a partner in the embryonic world of organic dairy. In our Northeast region, there were fewer than a dozen organic dairy farms at that time. The traditional farming support services, such as the U.S. Department of Agriculture and the universities, at best had little interest in organic and, at worst, condemned it.

We were in search of a partner, not just a supplier. Seeing farmers squeezed at every turn, we weren't keen to perpetuate what we saw as the failed model of conventional agriculture. We went looking for a partner with whom we could share a vision for change. We wanted a direct connection with the farmers and for them to feel connected to us so that we could learn from one another and problem-solve together. We sought to create something new—where farmers could be paid a fair price for their products. Where consumers could get a high quality, healthy product produced without harmful chemicals. Animals would be treated well and have access to fresh air and the outdoors. Farming practices would improve the soil, be safe for workers and not contaminate the environment. Who said the food system had to be about winners and losers? Why couldn't everyone, from farm to table, win?

* * *

In our quest for a partner, we heard about a little cooperative in Wisconsin that was expanding and forming pools of organic dairy farmers outside the Midwest. CROPP Cooperative. Admittedly we were skeptical: Was this just another group adding cost to the equation with little value? Were these idealistic hippies who, while well-intentioned, couldn't run a business?

We liked that they were a cooperative and that the management truly represented

the interests of the farmer-members. Straightforward and honest in their approach, they certainly knew a lot about dairy and organic. But CROPP was largely unknown in our region—and many dairy farmers were wary of cooperatives because of having been hurt by other co-ops. The people of CROPP, however, seemed so real and down to earth. There was no question that CROPP was a high integrity organization. So, after evaluating our options, we each took a leap of faith and decided to partner to grow organic dairy in the Northeast. Now we had to convince the farmers to join us. The road trips began.

Within a few years, thanks to some of our region's organic dairy pioneers—Jack Lazor of Butterworks Farm, Travis Forgues in Vermont, Gloria and Gregg Varney and Steve Russell in Maine, and the many other brave farmers who early on made the leap to organic and CROPP—there were thriving pools of CROPP farmers in Maine and Vermont. New Hampshire and New York farms were to follow. Today there are over 482 CROPP members in the Northeast producing 20 truckloads of milk every day.

Those early days mark the start of one of the most robust, healthy and important partnerships in Stonyfield's history. Together, we have innovated on cost structure and have combined efforts to educate buyers and the public on organic. We've mutually helped to fund farmers' transitions to organic, as well as provided education and training programs to help farmers improve their practices. The partnership is rooted in common values, shared vision and a deep commitment to the principles of organic. Together we strive for a win/win for all organic stakeholders—from farm to table. Together we have shared risk. Together we have learned from one another and grown as businesses and as individuals.

We've come far from those early days of farm meetings when we were looking for less than one truckload of milk per day. Today we acquire from CROPP family farmers 5 tankers worth of milk daily. Now there's a professional team of CROPP employees who are driving in the wee hours of the morning to the farm meetings around the country—and the shared values are still there.

At the heart of the CROPP/Stonyfield partnership is a deep mutual respect and open and transparent communications. We know that like organic, a partnership is a process of continual improvement—not a static or fixed place. Neither of us is perfect, and there have been some very rough patches, but together we strive for continual improvement. We have an enduring and profound respect for CROPP—what it stands for, and the staff and farmers who make up the organization.

We are so honored and proud to be a partner of Organic Valley and the amazing farmers of CROPP Cooperative for more than 15 years.

{ Nancy Hirshberg is vice president of strategic initiatives for Stonyfield Farm, the world's leading organic yogurt producer. For nearly 20 years she was at the helm of Stonyfield Farm's efforts to improve its environmental performance and reduce its ecological footprint. }

6. Finding a Better Way

to Make Better Food

2001–2006

* Focus * Sound business * New ventures

oor, okay, super * Roughly right or exactly wrong *

"I'm a living sunset
Lightening in my bones
Push me to the edge
But my will is stone . . .
'Cause I believe in a better way."
Ben Harper, "Better Way" (2006)

It was 2001 in tiny La Farge, Wisconsin, when George Siemon wrote about the Cooperative's 14-year journey. He thought back to 1987.

"We were frustrated," Siemon wrote. "The trends in agriculture were not in favor of small family farmers like us. We believed our customers were suffering worst of all. We had to find a better way to make better food."

By 2001, American agriculture had consolidated to unprecedented levels. Four companies controlled more than 89 percent of the cereal market, four meatpackers controlled nearly 80 percent of cattle processing, four companies dominated 49 percent of the broiler industry and one directed 30 percent of the dairy industry.

With America's food supply in the hands of so few, conventional farmers had fewer options for selling their crops, and commodity prices dropped lower and lower. They had to take whatever price was offered.

As "Big Ag" became more powerful, the more its dominance bolstered the organic alternative. For organic farmers, the market was growing, and they were being rewarded for breaking with agribusiness norms.

"We set out to provide our customers with the most wholesome, best-tasting, highest-quality food available anywhere, and we're doing this while protecting our soil, our water and our animals from unsettling trends," Siemon wrote. "We have eliminated all use of herbicides, pesticides, antibiotics and hormones. Our animals graze daily in the sunshine. The soil and streams are living again and they will continue to thrive under our care."

"To me organic is back to the way God made it, it's the instinctive way—to farm with nature and not against it, to make the best use of it and not fight it." Charlene Stoller, Ohio farmer owner

By 2001, just one percent of America's food supply was grown using organic methods, but that tiny percentage held promise. It was the culmination of six consecutive years of 20 percent industry growth. Organic was no longer relegated to natural foods stores alone. In fact, consumers found organic foods in most major supermarkets, at farmers markets, through direct-marketing Community Supported Agriculture (CSA) and even at some restaurants.

The pragmatic idealists at CROPP had reason to be optimistic. Organic consumers weren't just aging hippies or devout vegetarians. They were pregnant women and moms who awakened to the benefits of organic milk for their children's sake. They were health-conscious folks who jogged and juiced their way to organic food. They were seniors who thought that organic milk tasted just like the milk they drank as kids. They were aging Baby Boomers who believed they deserved the best food that money could buy.

By the early 2000s, mainstream consumers gave organic a hefty boost. *Reuters Financial News* reported American organic sales reaching $11 billion in 2002 and a projected $13 billion by 2003. CROPP's success actually preceded public awareness and the popularity of organic foods: in 2001, the Cooperative's sales totaled nearly $100 million, and by 2006, they had tripled to $333 million. CROPP had correctly anticipated growing

← James Frantzen with his farrowing sows on the family's farm in New Hampton, Iowa consumer demand and built its "bench strength" with enough organic farmers to meet market needs.

By the early 2000s, more than 60 percent of CROPP's grocery customers were mass market retailers including Fred Meyer and Publix, while only a few years earlier, the "big boys" of the grocery world represented only 30 percent of total sales.

> "Organic farming was born on the fringes of agricultural practice, but as its central ideas matured, it steadily moved into the mainstream." Ronald Jager, *The Fate of Family Farming* (2004)

"What we have here is one incredible grassroots movement," Siemon wrote in 2001. "Farmers and consumers are working together to make a difference in our world." The organic movement was a bright spot in the new century. More and more people believed their food choices had a lot to do with health and a better life.

The organic movement had become a revolution. It was the fastest-growing food segment in America, and plain-spoken Wayne Peters was amazed. "I didn't begin my association with CROPP thinking I'd be a revolutionary," Peters said. "I was just looking for a way of farming and marketing that made more sense, something that would allow my children to follow in my footsteps if they chose to." They have.

At Long Last, National Standards

The revolution spawned the first national organic standards in America. Creating them took nearly 12 years of research, decision-making, often-lively debate and blizzards of letters from consumers and activists. It all culminated in enactment of the National Organic Program (NOP) in 2002.

Looking back, the work started in 1990 when Senator Patrick Leahy (Vermont) and Representative Peter De Fazio (Oregon) sponsored passage of the Organic Foods Production Act (OFPA). As it turned out, a bright, young staffer named Kathleen Merrigan worked on Senator Leahy's staff and she helped draft the successful legislation. (She would later become Deputy Secretary of the USDA in 2009.) Other legislators before Leahy and De Fazio had advocated national organic standards, but America's Alar-tainted apple scare in 1989 gave the Leahy/De Fazio bill urgency. More surprises during that same era made consumers wary: some watermelons were laced with a pesticide called Aldicarb, a load of Chilean grapes carried cyanide residues. Then there was the "carrot caper," involving a conventional produce buyer who decided he could make bigger bucks if he simply labeled his conventionally-grown carrots organic and sold them at a higher price.

"How do I tell if something I buy is truly organic?" consumers wondered with good cause. News outlets offered guidelines: Yes, there are more than 60 organic

← Milk packaging includes the new National Organic Program/USDA Organic seal.

Washington farmers Kristie and Mark Schmid at the 2001 Natural Foods Expo in Anaheim, California →

Rooted in Citrus

The late Bert Roper, a third generation farmer, came home from serving in the Navy during World War II thinking about new ways to grow oranges in Central Florida. Roper's ancestors had settled in Florida in 1859 when the state was mostly cypress trees, scrub pines and lakes. At first, his family grew cotton and raised cattle. Then in 1917, Bert's father and uncle founded Roper Brothers, Inc., to grow oranges.

"In those days, Central Florida had about 150 small citrus operations, a lot like the small, family-owned dairy farms in Wisconsin and Minnesota," he said.

Bert Roper worked in the family business as a young man and earned his degree in plant physiology from the University of Pittsburgh. He focused his study on perfecting a process for concentrating fresh orange juice. With two orange juice concentrating plants, the Roper's business weathered tough weather cycles and volatile citrus markets as concentrated orange juice became an American favorite. Remember the advertising slogan of the 1970s and 1980s? "A day without orange juice is like a day without sunshine."

Concentrated juice eclipsed the fresh fruit business, and when Minnesota-based Cargill moved into Florida to launch citrus processing in the early 1990s, smaller growers saw the pervasive impact of corporate farming. "Cargill wasn't a grower," says Bert Roper. "Cargill was producing a commodity. We decided that we needed a niche market. I had studied the effects of chemical sprays used in orange groves and the dangers of fertilizers on the environment. For a long time, I wanted to get into organic."

Roper experimented with using fewer chemicals, slowly shifting to natural plant foods and minerals. He became a certified organic grower and produced his first organic harvest in 1998. The Ropers joined CROPP's organic juice pool when it formed in 2001. "We haven't suffered the yo-yo effect of conventional juice pricing," says Charlie Roper, Bert's son who now runs the business. "We have a stable pay price, and I believe more and more consumers are seeing the benefits of organic juice, even though it's more expensive than conventional."

It hasn't been all bountiful, perfect harvests for the Ropers since joining CROPP. In fact, Mother Nature's gut punches are a chief reason for the Ropers' loyalty to the Cooperative. When severe, unpredictable freezes and diseases harmed their citrus crops, the Ropers could not produce enough to meet demand. In other years, they faced an oversupply of juice. "CROPP stood by us during those times," says Bert Roper. "Because they are farmers themselves, they understand. They're not short-term thinkers, and they have big hearts. CROPP is the antithesis of a commodity trading company."

By "treating his trees like he treats himself," Charlie says his family's organic oranges have a higher percentage of fruit solids, and they taste better than conventional oranges. The wildlife around their groves has thrived, too. The Ropers see cranes and egrets wandering the nearby wetlands, a rare Florida panther prowling the edge of the groves, and, most evenings, two native alligators swim past the Roper family porch on Lake Butler.

← Citrus farmer and organic pioneer, the late Bert Roper; Bert with his son, Charlie, and grandson, Sutton

certifying agencies working around the country to ensure the integrity of organic production. But there are no national standards. Buyers beware.

The farming community, consumer groups and environmentalists joined forces, Katherine DiMatteo, executive director of the Organic Trade Association (OTA), said in 2002. In essence, they wanted to ensure that a farmer or food producer had to be certified to label their products organic.

When the OFPA was passed as part of the 1990 Farm Bill, it gave the USDA responsibility for the National Organic Program (NOP). The Act established a 15-member National Organic Standards Board (NOSB)—a group made up of environmentalists, scientists, organic farmers, food handlers and consumer advocates who had a key role in drafting the standards. It was a long and sometimes tortured process. As bold national initiatives go, another moon landing would have been easier.

George Siemon was asked to serve on the NOSB, and he co-authored the organic livestock standards. He also led a pasture task force that focused on strengthening the "access to pasture" requirements in the organic standards. It wasn't enough to give "access" to cows so they could "sometimes visit" the outdoors, he said. That wording created a potential loophole big enough for Bossy to jump through. Organic consumers expected their cows to graze on pasture whenever possible, Siemon said, and today they do, thanks to strict NOP pasture standards. Consumers have learned what organic farmers have known for decades: grass-fed cows produce milk that is healthier for humans.

Clovis Siemon, George's son, remembers seeing his dad in the family living room during those early years of frequent trips to Washington, D.C. "He had his head in his hands," young Siemon says. "It was obvious he was out of energy and over-burdened. With this new responsibility, he told us he'd be gone from our family even more. He wanted a weekend a month in D.C. like he wanted another hole in his head, but he wanted *real* organic standards even more."

The "Big Three" Backfire

Most consumers knew little about the machinations of writing these national standards, but many remember 1997 when a preliminary draft was circulated that proposed allowing "the big three" in organic production: sewage sludge, irradiation and genetically-engineered seeds (affecting a plant's natural DNA). The plan, openly opposed by NOP staffers, generated a USDA record 275,603 responses lambasting the big three inclusion. This groundswell demonstrated that regular consumers cared about the quality of their food and they knew how to mobilize for maximum impact.

A chastened USDA stepped back and sought a champion to continue guiding the controversial work, and Kathleen Merrigan, who had served as head of the Agricultural Marketing Services (AMS) in the Clinton administration, took the job. When the NOP was ratified in October, 2002, most organic farmers and clean food activists said the standards—though imperfect—retained the original character of organic production and were the strictest in the world. Many agreed, without the private certifying agencies serving as pioneers, there would be no national standards at all.

The arduous, 12-year process was probably beneficial, too. As organic food production matured during that time, farmers and organic advocates were more experienced, organized and in a better position to influence the standards' content.

At the same time, CROPP was already developing its own standards and sharing them with the industry. In many cases, CROPP's standards exceeded the national ones.

In 2000, Americans consumed 195 pounds of meat per person, 57 pounds more than the 1950s.

For the first time, consumers could look at produce or a food package and—if they saw the NOP/USDA stamp—know that the product was at least 95 percent organic. Foods with 70 to 95 percent organic ingredients could say so on the label ("made with organic milk," for example), but they could not display the organic seal. Makers of products with up to 70 percent organic ingredients could list those ingredients on the package, but they couldn't make other organic claims. Imported products had to meet the same standards.

Today, when the USDA gets countless calls from consumers asking, "How can I know a product is truly organic?" their standard answer goes like this: "Buy the foods with the organic seal. It's the only way to know for sure."

That coveted seal started to appear on more foods than most people imagined. Along with fruits, vegetables and milk products, it showed up on yogurt, juice, eggs, processed and frozen foods and—at long last—meat.

"Organic" Is Off Limits

Though CROPP had started a new organic meat pool in the 1990s, they weren't allowed to use the "O" word on their packaging until 2002, when the national organic standards were ratified. CROPP's "Valley's Finest" branded meat included beef (in 1996) and pork and poultry (1998), but before 2002, the meat pool seemed to operate in a kind of retail limbo, says Pam Saunders, the original manager of CROPP's meat pool. "Dairy, grains and other foods could use organic in their brand names, but we weren't allowed to because meat was under the jurisdiction of the USDA. All the other foods, including milk, were under the Food and Drug Administration."

Before 2002, CROPP *could* put "produced without synthetic hormones and antibiotics" on their meat labels, along with the name of the organic certifying agency. CROPP rejected using the word "natural" on its meat, because that word was by-and-large unregulated and meaningless. By this time, some so-called "natural" foods were actually genetically-modified. (To this day, there are precious few production restrictions for "natural" meat.)

CROPP had lobbied hard in Washington for more meaningful meat labeling, and they made their case in a particularly edgy poster. In a take-off on the famous "American Gothic" painting by Grant Wood, a sedate farm couple stands, pitchfork in hand, before their tidy farmhouse. Their mouths are gagged and the poster declares, "We're not allowed to tell you that our meat

One contented chicken on Gary Welsh's Iowa farm →

↑ Benedict Cook and his Minnesota livestock herd; James Frantzen, son of Tom and Irene, on their Iowa farm; Minnesota farmer Eric Miehlisch lets his turkey take center stage; CROPP wasn't allowed to use the word "organic" on meat packaging before 2002.

is organic." Pam Saunders handed the poster to Dan Glickman, U.S. Secretary of Agriculture, at a meeting in Eau Claire, Wisconsin, in 2001. It was a cheeky but effective gesture.

Still Focused?

CROPP's whole meat enterprise was iffy, financially. The venture tested the Cooperative's patience and called into question just how diverse CROPP could and should be. As CROPP diversified into more organic pools (adding orange and grapefruit juice in 2001, soy in 2004 and a grower pool supplying organic feed to members in 2008), some observers may have thought CROPP was losing its focus by doing too much.

Adding organic juices and soy offered synergy, however. These two new pools needed what already existed at CROPP: fluid processing plant partnerships, similar quality programs, a refrigerated distribution system and relationships with grocers who bought refrigerated goods like dairy. Starting a grower pool to produce organic feed became an absolute necessity in later years when feed was costly and in short supply.

Equally important, CROPP's mission was to serve as many organic family farmers as possible by providing them with markets for their production and sustainable pay prices. Yes, CROPP was focused. But meat was a whole different animal. It was an organic product, to be sure, but it required a different kind of focus than CROPP was used to: it involved different processing, there were different quality concerns, meat required frozen distribution (another new wrinkle for CROPP) and meat buyers were a totally different group. Some stalwart organic consumers were also wary of meat products because they had turned to natural foods decades earlier as vegetarians. Even some CROPP farmer-members were cautious. "What if Mad Cow disease comes to the United States?" they wondered.

In spite of these obstacles, meat was a natural fit for CROPP in other ways. Organic dairy and livestock farmers were certain that stress-free, unmedicated animals resulted in better and healthier food. In addition, when dairy farmers culled their cows because they could no longer produce milk or when the farmers had bull calves on their hands, they needed an organic plan for those animals.

> "Our meat is produced without herbicides, insecticides and hormones. The animals are raised humanely on small family farms by people who care about animal well-being."
> Pete Bassett, sales manager, Organic Prairie

"Why did we get into organic meat?" asks Louise Hemstead, CROPP's COO. "We got into meat because bull calves are born and most people want to eat meat. It's a natural outgrowth of the dairy business."

"The whole development of the meat pool was based on farmer demand," says Saunders, a former organic farmer herself. "They wanted in." But starting, growing and sustaining the organic meat pool was like running an obstacle course. More than any other venture (except perhaps CROPP's own origin and evolution), it demonstrated a lesson that Paul Repetto of Horizon Organic had shared with CROPP years earlier. If you're going into a business, he said, you should have a plan and be willing to accept the results if the business does poorly, if it does OK, *or* if it's super. If you're comfortable with the risks and results of all three scenarios, you can afford to enter the business. It's dangerous to be too confident of success.

The Cooperative had also learned that business plans are necessary tools, but business assumptions can change frequently. CROPP's leaders joke that the wisest course in following a business plan is to divide the plan's projections in half and multiply by zero! This is CROPP's way of reminding itself to rely on the advice of others *and* their own intuitive sense, as well as on a business plan.

CROPP was confident that entering the organic meat business was the right thing to do, but that confidence did not guarantee financial success, as the meat pool's long slog toward profitability proved. The Cooperative became familiar with the "poor" and "OK" aspects of Paul Repetto's business scenario, but the "super" part would have to wait.

Patience Pays

The entire meat enterprise required great patience. The program was a prime example of CROPP's long-standing willingness to subsidize new pools and invest in them until they can pay their own way. "We've always been willing to invest in start-up programs," says George Siemon, "because we're building an infrastructure to serve the whole nation. We lost money on the meat business and the first question was, 'Do we give up or throw more good money after bad?' The real questions were, 'Do consumers need an organic meat market?' and 'Do farmers need an outlet for their livestock?' The answer is 'Yes' to both. Once farmers join our team, it's not fair to say, 'We're going to abandon you now.'"

> "We're not an organization that makes decisions purely on short-term profits."
> Eric Newman, vice president, sales

Subsidizing new pools depends on CROPP farmer-members being willing to make investments, even if they're not directly beneficial to their own pool. Siemon calls it "a community approach" that focuses on the long-term health of the whole enterprise. When CROPP

Quirky, with Integrity

When a company can't afford big, expensive marketing campaigns, they often resort to low-cost, unconventional tactics that require more creativity than cash. It's called "guerilla marketing," a term developed by marketer and author Jay Conrad Levinson. He linked unique, low-cost marketing methods to the atypical tactics of guerrilla warfare, and the term stuck.

The guerilla approach worked for CROPP in the fiercely competitive 2000s. There was no money for big, splashy campaigns. That kind of spending didn't fit the cooperative culture, anyway. Instead, CROPP relied on public relations, poster campaigns, internet marketing, partnerships with respected opinion-leaders, and a grassroots farmer ambassador program to spread the word about organic. CROPP created a buzz with consumers and retailers with an approach observers called "quirky, with integrity."

"We may give our cows acupuncture, but we draw the line at hot tubs." The poster depicted two udderly serene milk cows seated in a steaming tub out in a farm field. The message explained that humane animal treatment was a central element of organic production methods, including—if need be—milk cow acupuncture treatments for better health, instead of antibiotics.

In another poster—thanks to the photographic wizardry of CROPP's then creative director, Carrie Branovan—Robert Schmid, a handsome hunk of a farmer in Trout Lake, Washington, magically stood at ease on the glassy surface of a mid-summer lake wearing his ice skates. Nearby, his milk cows grazed. "It's all about balance," the poster copy declared. "Our family farmers never treat animals with artificial hormones or drugs and dairy cows always have plenty of pasture space. By farming organically and treating animals humanely, we're helping put our earth back in balance."

The brains behind these quirky with integrity messages were Theresa Marquez, CROPP's then chief marketing executive, Branovan and agency partner Rick Dalbey—folks with a formidable creativity quotient.

CROPP's in-house capabilities grew through the 2000s with an art department that created point of purchase displays, brochures, sell sheets, new packaging and trade booth wares. The new marketing communications department worked on trade and consumer events, print advertising, videos, billboards, radio spots and creative partnerships. They even began publishing *RootStock,* a magazine about the organic lifestyle created expressly for consumers by Branovan and staffer Andy Radtke. Readership grew. A friend of CROPP even reported seeing a tourist bus passenger in Cancun, Mexico, poring over it.

With the recrafting of the Organic Valley brand mark, by 2002 all OV branded products had a new, clean unified look. Half gallon milk cartons were "mini billboards" carrying messages about the benefits of organic food from the Children's Health and Environmental Coalition and activist Erin Brockovich. Later, nutrition expert Dr. Phil Landrigan of the Children's Center for Health and the

F FACTORY FARMS!

Environment, helped CROPP publish "A Month of Healthy Lunches," featuring organic choices.

All these efforts were intended to turn first-time organic customers into life-long, informed customers and friends of CROPP.

Without a doubt, CROPP's farmers made the strongest connection to consumers. In 2004, Theresa Marquez had a brief but powerful conversation with a retail customer. "You know what I like about Organic Valley?" the customer asked. "You're the real deal." That set Marquez thinking. The "real deal" of CROPP had always been its farmers. The people who created the Cooperative. The people who understood organic farming best.

Willing farmers who participate in CROPP's Farmer Ambassador Program help spread the word about organic foods, their health benefits, and how organic practices restore the environment and ensure a decent livelihood for family farmers. They appear at events, including GreenFest (environmental events held in major cities), Farm Aid, the Go Green Expo in New York City, Seattle's SeaFair and Chicago's Family Farm Expo. They appear at Adopt-A-Store events in their communities. They speak at gatherings for young farmers, at universities and primary and secondary schools. They are interviewed by journalists, they host tours on their farms, and they share their stories. They underscore the "know your food, know your farmer" theme, and some of them have gotten so good at it, they've become minor celebrities on the "green circuit."

Some of those farmers are also part of Generation Organic™ (Gen-O™), a program started in 2006 that highlights young people between 16 and 35. They have high-minded goals: to support one another with encouragement and wisdom, to mentor those who are considering a future in organic farming, and to demonstrate that organic farming is a noble, viable and desirable profession.

They have fun doing it, as anyone who has seen the Gen-O "magic bus" touring from coast to coast can attest. Parked in the heart of Washington, D.C. in 2010, the bus was a rolling billboard for organic with this compelling challenge: "Own your food, drive your future!" The young farmers even met with Deputy Secretary of Agriculture Kathleen Merrigan and shared their ideas with USDA Organic and Sustainable Agriculture Policy Advisor Mark Lipson. Later, they toured ten major East Coast colleges and universities.

The next year, 2011, the Gen-Os headed west to Montana, Washington, Oregon and California on their second "Who's Your Farmer" tour, and in 2012 they toured the Co-op's midwestern heartland. They had the full support of CROPP's farmer-members behind them, too. Since 2007, farmer-members have voluntarily contributed to Farmers Advocating for Organic (FAFO), a fund that supports, among other things, educational projects just like this.

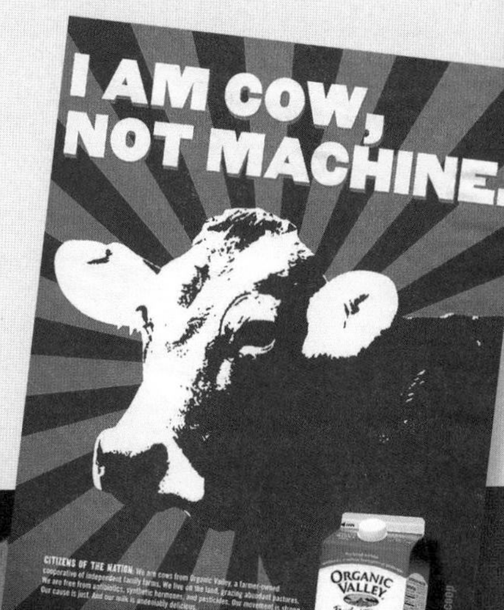

← As CROPP expanded across the country, more organic family farms, like the Deal farm in Mount Vernon, Texas, could count on a reliable livelihood.

Herman Roberts on his family's farm in Preston, Idaho →

expanded the geographic reach of its organic dairy production with two new pools in Colorado and Texas in 2005, for example, they needed start-up support just as several pools before them had.

These investments are not without discipline and accountability. CROPP created a system for tracking pools that are breaking even or losing money, and the Cooperative expects realistic plans for achieving self-sufficiency. All CROPP pools share the Cooperative's profits, but the pool that is losing money cannot reap these returns until it has repaid its losses.

CROPP's organic meat business would finally have its first profitable year in 2010, after 12 years of struggling to break even. It was a long time to wait, but by that year, Organic Prairie (CROPP's new brand name for its organic meat line) meats were sold in 50 states and ordered online from eager international consumers, especially in the Asia-Pacific region.

> "It would have been easy to say, 'Our dairy business is going great, why do we need the meat business?' Having multiple pools rounds out our Cooperative, and it includes the full cycle of the farm." Pete Bassett, Organic Prairie sales manager

A Sure Winner

When CROPP launched its Organic Logistics subsidiary in 2004, there was no reason to imagine poor or even OK results. It looked like a winner from the start.

> "Our product travels on milk trucks, over-the-road tankers, refrigerated trailers, and even trains." Louise Hemstead, CROPP chief operating officer

By the early 2000s, CROPP had learned to be agile by staying as close to its farmers as possible in regions all over the U.S., while delivering products to customers on time in locations just as distant and diverse. Why not turn these logistical skills into a business? "We were already helping smaller, organic folks get their products to customers, so we created Organic Logistics," says Louise Hemstead, CROPP's chief operations officer. "It was a natural outgrowth of our business." It was like sharing trucking services and know-how with a neighbor. Milk customers, including Nancy's Yogurt and Stonyfield Farm, could distribute their products by tapping into CROPP's reliable and cost-effective national network.

With more products to move, CROPP could negotiate better rates with carriers and warehouses. The new subsidiary helped CROPP reduce its transportation

← CROPP launched Organic Logistics in 2004; John Kolar is CEO.

CROPP's Cashton, Wisconsin, Distribution Center is a "one stop shop" for organic products. →

costs by selling its distribution services. Natural and organic food producers got better transportation and warehousing rates and fast, national service they could not achieve on their own, while CROPP reduced environmental impacts and costs by filling its trucks.

"We understand the importance of supporting the premiere organic brand with a distinguished level of support, and take great pride in delivering on time, as ordered, and in good condition." John Kolar, CEO, Organic Logistics

March 2003: Conventional dairy farmers get paid the lowest in more than 20 years: $0.78 per gallon.

Grab "The Big Ring"

They called it the organic dairy "gold rush" in 2004, and it continued until 2008 when America tumbled into the deepest and longest recession in modern history.

By 2004, the organic foods market had come of age. Whole Foods, started as a single store in 1985, in Austin, Texas, had gone public and was on its way to becoming a 310-store behemoth. Trader Joe's was popping up in major cities. Mass marketers wanted their share of "the big ring," too. This is retail-speak for organic shoppers whose food selections typically add up to a bigger bill at the cash register than those of non-organic shoppers. In their own research, Walmart discovered that some customers went to Whole Foods for their organic purchases and Walmart for the rest of their groceries, household and kitchen goods. "Let's grab more of the 'big ring,'" Walmart and other big retail players declared. As it turned out, Walmart was so good at seizing this new opportunity that they became the biggest seller of organic milk in the U.S.

"We have made a strategic decision to steer clear of any bidding contest with our competitors." George Siemon, 2004

CROPP had been selling its organic milk to Walmart since 2001, and it hadn't been an easy decision to make. Some of CROPP's employees saw Walmart as "the antithesis of everything their company stood for—an impersonal, price-slashing menace, a killer of small businesses and rural downtowns," *Inc. Magazine* reporter Kermit Pattison wrote in 2007. Some of CROPP's long-time natural foods customers asked them not to supply Walmart. Ultimately, though, CROPP's management team voted to go ahead because consumers were asking for organic, and Walmart made organic foods available in places they'd never been offered before.

It was a sound business decision. Early in its life,

Beneficial Bean

Maybe it was the intelligence work Wayne Wangsness did for the U.S. government during the Cuban missile crisis in the 1960s. When he came home to Decorah, Iowa, after serving his country, he applied his considerable problem-solving skills to soybeans.

To be specific, Wangsness and a group of innovative organic grain farmers in the northeastern quadrant of Iowa decided to make a better soy milk using all of the bean, not just part of it.

"We were looking for a way to add value to our soybeans," says Wangsness, "and we recruited partners who could be critical to our process." One was World Food Processing of Oskaloosa, Iowa, a company that had developed the technology to convert whole soybeans into soybean powder. "Conventional soy milk is made by grinding the beans a bit, soaking them and then squeezing out the liquid to make the milk," says Wangsness. "The rest of the bean is discarded. We figured that the soybean was made by the soybean plant in proportions that are healthy, and we should use the whole bean, not just the soluble part."

Their second partner was FTE Genetics, a sister company to World Food Processing. They created a soybean seed without genetic modification that was superior nutritionally and free of the bitter taste found in conventional soybeans.

Wangsness and about 25 grain farmers formed the Quality Organic Produce Cooperative (QOPC) in 2000. It took them years to invent their better soymilk and sell CROPP on starting an organic soy pool. More than a few dairy producers looked askance at soy. Wasn't it competition for cow's milk? Was it *real* milk? In the end, CROPP's dairy producers greenlighted the project, but requested that the product be called "soy beverage" or simply "Soy."

While the farmer-members of QOPC worked through five years of product development, they sold their organic soybeans on the open market and watched the prices rise and fall. That uncertainty alone gave the farmers a reason to find a better way, says Wangsness.

He studied world health statistics and discovered that a surprisingly high percentage of people struggle with lactose intolerance (an inability to digest sugars found in milk and dairy products). He also followed the accumulating scientific evidence that confirmed soy's disease prevention benefits, attributed to the consumption of isoflavones found in soy. *The New England Journal of Medicine* reported in 1995 that soy consumption reduced cholesterol levels significantly and the Nutrition Committee of the American Heart Association in 2000 reported that 25 grams or more of soy protein in a daily diet promotes heart health.

CROPP started its soy pool with Wangsness and 14 Iowa organic farmers in 2004. This new partnership gave them a stronger link to organic grain farmers, the allies of organic livestock producers. And the farmers produced one terrific bean. "Our soybean variety and our process is unique to us," Wangsness says. "We've even worked out a system of processing so you can trace your soy milk to the specific farmer who grew the soybeans in your carton. Not only is that a unique personal touch, but it's a great way to track product quality."

CROPP had learned to pay close attention to what customers wanted, and that included grocery retailers. If consumers wanted an eight-ounce organic cheese, CROPP produced it. If they wanted calcium in their organic orange juice, the Cooperative provided it. If they wanted to buy their organic foods at big chains, CROPP geared up and supplied them. If CROPP was to succeed as a mission-based cooperative, it had to run a sound business with good business practices and good customer service.

The decision to supply Walmart reminded CROPP of earlier days when Wayne Peters declared that if the tiny Cooperative didn't dare play outside its league by putting the quality of its cheese up against established cheesemakers, it would never reach the big leagues. Now, CROPP was playing in the big leagues of mega-customer orders. Walmart became CROPP's third largest customer and perhaps the most rigorous of them all. On-time delivery meant more than never being late. Suppliers got marked down on "report cards" for being minutes early, too.

> "Organic Valley has remained independent as other organic pioneers have been snapped up by the Jolly Green Giants of the food industry."
> *Inc. Magazine*, July 2007

New Alliances and Bidding Wars

All that new demand for organic milk in the mid-2000s, especially among "the bigs" of retailing, created the organic milk gold rush. "If you had milk supply, you could sell it, so a lot of people got into organic milk production," says George Siemon. "All they had to do was sign up dairy farmers."

H.P. Hood Dairy jumped into the organic fluid milk market by partnering with Stonyfield Farm. Hood was aggressive in attracting organic farmer producers. The company offered one-time signing bonuses to organic farmers, igniting a bidding war for organic milk. At first, this seemed like good news for farmers as organic pay prices spiked to an all-time high in 2004.

Being the "Price Keeper"

The new competition put CROPP in unfamiliar pastures. "We were no longer the price leader for farmers," Siemon says. Instead, CROPP became the "price keeper" as organic dairy prices became volatile.

How long would this new era of supply and demand pricing for organic dairy continue? CROPP's goal had always been steadily increasing pay prices that were stable and sustainable, but this new market was anything but stable *or* sustainable. CROPP made a strategic decision to sit out any bidding contests with its competitors. "We don't want to see pay prices go

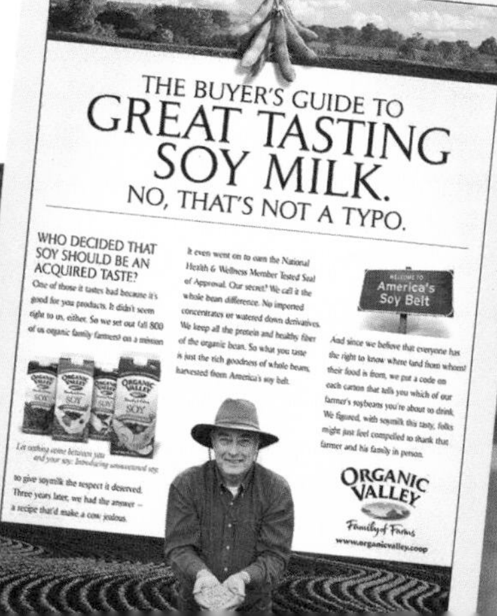

← Organic soy farmers (from left): Wayne Wangsness, Todd Banes, Ryan Wangsness, Erwin Henderson, Todd Boss, Paul Hunter and Dan Parizek; Wayne Wangsness with his granddaughter, Savannah

up, only to fall again in the future, even though we now find our farmer recruitment efforts outbid by corporate competitors," Siemon said in 2004. "We believe it is more important for us to be the 'price keeper,' and make sure that our pay program doesn't slip into the conventional mode of constant ups and downs."

By 2004, the size of CROPP (689 farmer-members) and the strength of its Organic Valley brand gave them the heft to take this stand in the marketplace.

CROPP also eventually took a stand that some observers thought was business suicide. It happened because the "gold rush" predictably began to crash, leading to "dry Thursday," when there simply was not enough organic milk to go around—not even from CROPP family farms. In December, 2004, CROPP's leaders gathered to review all 200 of their dairy customer accounts with an eye on who would get milk and who wouldn't. Would CROPP favor the many, smaller operations that helped them grow (about 45 percent of the Cooperative's sales), or would they choose the bigger accounts with huge growth potential? By 2004, Walmart was buying about 1.3 million gallons of organic milk from CROPP each year.

As CROPP worked through its customer list that day, deciding whom they would supply and whom they wouldn't, they considered Walmart. If Walmart became too important to CROPP, it could consume so much of the Cooperative's milk that CROPP could become vulnerable. Walmart could also press for lower prices, damaging CROPP's commitment to stable, fair prices for farmers. While Walmart hadn't asked for price cuts, a major rival had already offered their organic milk at 15 cents less per half gallon.

Ultimately, CROPP largely chose the natural food stores over the mass-market accounts, and that's where CROPP's notion of "roughly right or exactly wrong" came in. "We try to remind ourselves of the importance of balancing raw data with gut instinct," Siemon explains. "Some have said that employees with advanced degrees often struggle with working at CROPP because we don't go at business in the same way the textbooks advocate. Chasing the last decimal point in a plan could be the most important fact . . . or it could be irrelevant. Telling the difference is the real challenge."

In December 2003, 53 countries banned the import of American beef due to a mad cow disease scare.

"We're seriously intent on doing the right thing and deciding what is best for our mission and our business."
Cecil Wright, director of sustainability and local operations

CROPP decided to stop shipping organic milk to Walmart on January 1, 2005. In all, CROPP stopped selling its organic milk to about 15 of its 200 customer accounts including several supermarket chains. While most customers came back to CROPP after the milk supply issues were solved (Walmart included), some did not. "Walmart was never negative to us," Siemon told *Inc. Magazine* in July, 2007. "We just made a business decision about our best odds of longevity." And it was more than "roughly right." In 2005, CROPP's dairy sales increased by 15 percent and sales grew again by 37 percent in 2006.

When the organic milk shortages of the 2004 "gold rush" began shifting to oversupply (even before the 2008 recession), some of the newest players in the organic market crashed and exited. CROPP had lived through plenty of over- and under-supply periods in its 17 years. They had learned that once supplies were short, it was too easy to lose track of real demand, sending prices into chaos. "We must always remember that organic food is a partnership between the organic farmer and the organic consumer," Siemon cautioned. "We cannot let short term conditions force us to push our price point too high at the retail level."

CROPP: Rooted in Rural Life

If CROPP wanted to "chase the last decimal point" in its plan for growth, it surely would have left little La Farge, Wisconsin, when it needed a new home. CROPP was on its way to employing more people than the town's own population of 750.

Once again, the Cooperative took an unconventional path that seemed absolutely right. "La Farge was the town that grew us up," says Siemon. "We didn't want to be like a regular corporation and leave." The Cooperative had, from its origin, believed in supporting rural communities. CROPP's employees, numbering about 300 in 2004, lived either in La Farge, Viroqua or other nearby towns. CROPP's leadership didn't want their lives disrupted.

The perfect location, covering about 34 acres up on a hill just north of the La Farge Village Park and neighboring the scenic Kickapoo Valley Reserve, was not for sale. The widow of farmer John Stout and her son were happy with their land and their home overlooking the acreage that John farmed for years.

"We looked all around and I kept saying, the hill's the only place to be. It's a beautiful location," George Siemon says. He called on the widow Stout and her son.

"We want to stay in La Farge," Siemon told them, "but we need a place to land. I believe CROPP was meant to be on this hill. We've been growing fast. Staying here will be good for the community."

This Letter

I received this letter out of the blue in the summer of 2002 from an organic producer in transition located 75 miles north of me, (thus the Google map on the back), the milk inspector mentioned to Joe Hammon that he knew some one interested in organic and he tracked me down. This chance letter turned into a friendship that lasts to today, Joe and I have become confidants and talk often, (our wives say too often).

From this chance contact we started to build a pool, Joe was so aggravated that he could see an OV transport on the 4 lane near him almost daily going to Wisconsin from Amish country in eastern Ohio and they would not stop. As you can see from the doodling on the letter, over the next 2 years we plotted and attended meetings, most 2 to 3 hours drive away and built a pool of sorts,

In the middle of this whole mess I suffered a loss of a son and the subsequent heart attack of my Father and partner, honestly much is a haze for about 16 months from that point but honestly that letter and the affairs that it sent in motion saved my life both financially and spiritually, with out the efforts of Joe, and many other farmers and the work of Jim Wedeburg of OV, My Dad and I would probably have sold the cows and quit, ending a 5 generation tradition. Now thanks to OV the 6th generation has a future

All due to one short letter.

Thank You, Joe

David Osterloh
Maria Stein Ohio

2002

Hi,
I am an organic farmer and am less than a year away from my dairy being certified. Our milk inspector told me you are interested in selling organic milk. The last time I talked to the organic buyers of milk, they told me if a load could be put together they would come the distance. I am very interested in your location and time frame. No matter if you are close to certifying or just contemplating. Please contact me. Hopefully we can help one another to move to a common goal.

Sincerely,
Joe Hammon
10607 Wonderly Road
Mark Center Ohio 43536

Organic Valley

18.25

Bob Felkhuster
Stan Gamby
Gamby
17717 St Rt 31
mt Victory
43340

6th at Paul's 7:30

Lagrange - 8:30 - 12:30 wed 6th
Cropp intro

Food is Love

Days after Hurricane Katrina decimated the Gulf coast on August 29, 2005, Clovis Siemon and Felipe Chavez drove two buses loaded with organic food, a portable kitchen and Wisconsin volunteers toward Mississippi.

"I knew where to find a good bus. I knew my barn had enough kitchen equipment to feed thousands of people. So I thought, 'I'm *response-able.*'" Robert Clovis Siemon

Never before had anyone from CROPP transported a rolling kitchen and volunteers to a scene of indescribable loss. The whole venture was a spontaneous gesture without any of the usual planning, permits or approvals from the Red Cross, FEMA, local security or police.

Since its founding in 1988, CROPP has encouraged employees to volunteer for causes they care about. The Cooperative has also contributed millions to organizations that protect family farms, communities and the environment. Some of the major recipients of CROPP's giving are the Children's Health Environmental Coalition, Farm Aid, Organic Farmers Research Foundation, Bioneers, the Waterkeeper Alliance and Heifer International.

The Katrina disaster required a different kind of giving. It was unplanned and risky.

The Organic Valley two-bus caravan stopped for gas on the outskirts of Mississippi. A local man approached. "I know where you can serve food," he said. "My community is in dire need. Stick with me and I'll get you past the barriers and armed security."

It was late as the buses approached the devastation and the volunteers could only see what the headlights saw: crumbled houses, overturned cars, dead animals and finally a pickup truck hanging by its axle from the top of a telephone pole 30 feet in the air. They all gasped and fell silent, realizing how high the storm surge had been. Katrina had been one of the five deadliest hurricanes in U.S. history.

The New Waveland Café set up for serving on September 12, 2005, in the tiny town of Waveland, Mississippi, population about 6,600. It had been nearly destroyed by Katrina. The Café wasn't on anybody's formal list of volunteers, but when questions arose about their right to be there, the Waveland Police Department became their champions.

"The next day, we woke up and started cooking breakfast: eggs, bacon, the works," says Clovis Siemon. About 150 people came to eat, and the number doubled by lunchtime. By dinner, about 400 people waited for food, and at lunch the next day, 2,000 came.

As word spread and several weeks passed, the New Waveland Café was serving about 4,500 meals daily and quickly running out of food. About 30 rotating volunteers from all over the U.S., including a steady stream from CROPP, prepared and served meals. Along with additional food shipments from CROPP and Sanderson Farms

(chicken), churches pitched in with donations of food and supplies, filling the gaps. Every day, says Siemon, "we asked for help with supplies in the morning, and by midday we would find a pastor somewhere—Texas, Tennessee, Wisconsin. They'd collect what we needed, and more. Someone would drive all night and arrive with a trailer full of supplies by the very next morning!

"It was a daily miracle," says Siemon. "CROPP stocked us up, but human giving kept us afloat."

The Café operated for three months following the aftermath of Katrina and through a second hurricane, Rita. "When it looked like Rita would reach land, virtually every relief kitchen packed up to go and the Red Cross was pulling out," says Siemon. "We talked about it, flipped a coin and decided to stay, even though Rita was projected to be twice as strong as Katrina. Everyone was telling us to leave, but not the local people. They had been wiped out and they weren't leaving. Every day, they came to eat. We tore down our kitchen and set it up in a smaller form. We buckled down and boarded up our Organic Valley bus to protect it. We stayed and cooked. "

Just before Rita hit, Siemon got a call from his grandmother Margaret, a Floridian familiar with the force of hurricanes. "Clovis, I'm really worried," she said. Her grandson immediately assumed she was concerned for his safety.

"Clovis," she said. "You're cooking in the south. Are you using bacon grease? You need to use bacon grease on everything. It's important that you cook southern!"

They were, but they added an extra dollop after Margaret Siemon's emergency call.

Since Hurricane Katrina, CROPP has sent the food kitchen bus to other disaster regions, including hurricane Sandy in 2012.

(Left to right): Organic Valley's bus carried volunteers and cooking equipment to the Gulf Coast; it was later boarded up and tied to an escape boat in anticipation of Hurricane Rita's strike; Lumen Hobbins, a CROPP community member, serves salad, Spanish rice, fried chicken and smoked pork chops—almost all of it organic.

They warmed to the idea.

"Your husband really helped people like me when we moved to the area and we wanted to learn about farming with horses," Siemon said. Stout didn't unfairly judge the back-to-the-land novice farmers with all their idealism and energy. He helped them.

The Stouts sold CROPP their land and, in August, 2004, construction of CROPP's new headquarters was complete. One hundred eighty employees moved into the three-story structure (designed to be responsibly "green") with a distinctive barn facade and lobby. The headquarters was positioned on an east-west axis to make the most of northern and southern sun exposure through walls of non-glare windows. The ceilings, walls and foundations were insulated to high R-values and the insulation itself was recycled material. An Energy Recovery Ventilator reduced energy consumption by capturing the energy of the existing air in the building to either heat or cool the incoming air. This allowed the building to have plentiful, healthy fresh air while still saving energy.

"Our insulation is made from old blue jeans and there are 33 miles of computer and telephone cable and eight miles of electrical wires in the building and one cow weather vane atop the cupola." Cecil Wright, describing CROPP's new headquarters

Steel used in the building's siding was recycled. The interior wood trim came mainly from local or sustainable sources. The parking lot lights had stand-alone photovoltaic systems. Nearly all the waste generated during construction was recycled.

Settling into their new workspaces, CROPP employees found fewer walls and more space for interaction and collaboration, lots of natural light and a café serving organic meals daily.

They were finally all together in a brand new building created expressly for CROPP and rooted in its birthplace.

↑ When CROPP moved into its new home in 2004, the staff gathered for a celebratory group picture.

"Croppies" had dinner on the acreage that would soon become the Cooperative's new home; CROPP's new headquarters overlooks the Kickapoo Valley Reserve. →

Soul Food from a Farm Girl * *Sarah Holm*

Finished with my garden, I picked up my basket filled with organic dill and took my little brother by the hand. As I tasted a piece of the delicate leaves, I reflected on how fortunate I was to have a farm and garden where I can raise my own ingredients. Then, after considering our culture of convenience, I saw how blessed I am to simply know how to cook.

I suppose women became the food-makers because it is simply the continuation of their role as life-givers. A woman gives life through birth, then nurtures it with her own body, then continues to sustain it by putting dinner on the table.

I do not want to discredit the men who farm. I am very glad for my father's help and would not want to farm without him, but it is simply a fact that women are farmers, too. I should know—I farm with my mother and five younger sisters.

It is encouraging for me to see Organic Valley supporting women farmers, because in the past few decades our traditional role in "putting dinner on the table" has been demeaned and nearly lost. Working in the fields, gardening, canning, and cooking dinner day in and day out is too often looked at as lowly work. There is no doubt that it is hard work, but this responsibility does not chain a woman—it gives her power.

The decisions a woman makes about how and what she raises and grows, and how and what she feeds herself and those she loves reverberates through everything. Sometimes changing the world starts at home with a garden and a home-cooked meal.

My peers sometimes ask me how I can stand to eat animals I have nurtured from birth. It's true that when I started farming, this was hard for me, but now I struggle to eat animals I have not raised. What had the animal itself eaten? Did it have a good life? I personally feel it would be cutting corners to purchase meat without a name, neatly cut into pieces and packaged to disguise its animal-ness. I could not think myself compassionate for doing so.

One day my little sister came to me with a hurt chicken. It was quite young, having grown its adult feathers only a week earlier. I held the chicken and stroked its head. There was nothing I could do for her. I was sad that she would die so young, but farming has taught me that death is a part of life. The real crime against nature is not death, but waste. I killed her, cleaned the carcass, and began cooking it before I had even had breakfast. That morning I began to comprehend the responsibility of the farm women who pursue with tenacity their task of feeding the world a few people at a time.

If you love someone, give them food that nourishes their body and soul, food that is grown responsibly, and food that benefits the world. May women everywhere realize that they have been entrusted to feed the world—not only its belly, but its very soul.

{ Sarah Holm farms with her family in Elk Mound, Wisconsin, on their Organic Valley dairy. She loves to farm, garden, and write. She has a degree in Political Science from UW-Eau Claire. }

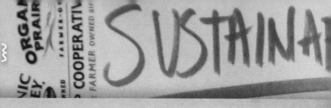

Reflections on CROPP's Core Strengths

** Chuck Benbrook*

As the organic food sector has grown into a real industry over the last few decades, I believe there are two differing viewpoints on the core purpose of the organic food industry.

These viewpoints boil down to the following: The organic farming community is intended to

- *Save the family farm, and by doing so, create a socially just, sustainable alternative to the rest of the food system, or*
- *Provide healthier food and feed for people and animals, thereby promoting their health and the health of the planet.*

Since its origin, I believe that CROPP has gone a very long way in demonstrating how these two viewpoints can truly complement each other.

Wisdom is not like water, flowing downhill first in small streams and then into rivers that grow large and strong as they reach the sea. Wisdom arises erratically and is often not even recognized for what it is. It does not reproduce or spread spontaneously. Without setting out to do so, CROPP has invented a way to capture and share wisdom that is greater than the sum of its parts.

As the publicly funded, university-based cooperative extension system has withered on the vine, lost its farmer-first focus, and become beholden to companies with commercial interests to defend and promote, CROPP has created a viable alternative that demonstrates, day in and day out, the value of investing in the creation and sharing of collective knowledge.

CROPP's unique strength arises from its willingness to confront problems when they arise by first coming to understand their root causes, and then assessing how to make those root causes go away, or at least become manageable isolated events. This inclination to use careful analysis and facts to tackle problems directly reflects how organic dairy farmers can become successful and sustainable. It is also the surest path to a healthier planet, allowing the organic community to find a way for advocates of all stripes to lend their energy in positive ways to make organic food both better and more common.

Back in the day when CROPP was born, organic dairy pioneers were motivated by their observations and understanding of what was going on with their land and the health of their animals, and sometimes themselves, their families, and neighbors. There was not a lot of science comparing organic and conventional dairy systems, and indeed, it was difficult to actually define what an organic dairy farm was.

As CROPP grew, it gained strength in multiple dimensions. First and most important is the diversity and depth of skills and experience across the CROPP farmer pool. CROPP invested heavily in the skills and technical acumen of its farmers. CROPP leaders recognized that no one had all the answers. In fact, there was far too much folklore, snake oil, anecdotal observation, and wishful thinking circulating as proven fact.

The farmer-leaders of CROPP understood that it takes years of experience and support from others with unique skills and insights to learn how to run an organic dairy farm that consistently has healthy ani-

mals, produces high-quality milk, builds soil quality, and delivers profits sufficient to support the farm, the animals, and the family and workers.

But CROPP leaders have come to be properly skeptical when scientists—who have spent their careers working within conventional dairy management systems and science paradigms—offer opinions on what is happening on organic dairy farms and what farmers need to do to overcome some problem.

As I have gotten to know CROPP farmers and CROPP staff scientists and technical specialists, this stands out: CROPP's vets and animal scientists learn as much from their farmers as they did from their university training, professional networks and past experience. Over the years, these professionals have become melting pots of collective wisdom, pieced together from the observations, questions, insights, experiments, failures and breakthroughs that have played out on hundreds of CROPP farms.

In addition to supporting their skilled farmer-experts, CROPP was able to make investments in the infrastructure needed to more efficiently and safely bring milk from its farmers to the people wanting to buy it. CROPP's distribution system has come a long, long way. While some might not appreciate the compliment, CROPP has come closest to Walmart-scale efficiency in its distribution.

CROPP is also willing to continuously question and rigorously test the basis of its belief—and marketing messages—about the human and animal health benefits of organic dairy farming. Over the years CROPP has demonstrated that it recognizes the need for continuous improvement and how to achieve it. CROPP also understands that the new customers of tomorrow are going to need better data and more convincing analyses of the benefits of organic to change their behaviors and purchasing patterns.

CROPP has invested in rigorously testing its milk quality and safety, and it has supported development of new tools to quantify the performance of organic dairy farms and the nutritional quality of organic dairy products. CROPP has a positive, powerful story to tell about the omega fatty acid benefits of its milk and dairy products, and about cow health and longevity. They insist on identifying quantifiable and meaningful benefits that improve people's lives and the lives of the animals that sustain us.

CROPP, by its example, has demonstrated how two viewpoints on the goal of organic farming can be successfully merged. CROPP has both saved family farms and provided healthier food.

{ Dr. Charles Benbrook is a research professor at the Center for Sustaining Agriculture and Natural Resources, Washington State University, where he is leading the "Measure to Manage: Farm and Food Diagnostics for Sustainability and Health" program. In cooperation with several OV farmers he created the "Shades of Green" dairy farm environment footprint calculator. }

Threads of Wisdom

Discoveries along the journey

Threads of Wisdom

by George Siemon

The beauty of the experience of CROPP Cooperative is how much pioneering we did without realizing how leading edge we were. The following 'Lessons Learned' are just a start to what we have learned along our way. We hope that they are as meaningful to you as they have been to us.

MISSION

* *Positive mission-based business*
CROPP Cooperative has been blessed with a clear mission that has steadily guided us. It's a hopeful vision to embrace organic farming and enable the sustainability of family farm agriculture. This sustainability starts with the care of the Earth but also has a human focus; the sustainability of both the business and the family farmers we serve. *(pages 33, 175, 195)*

* *Y in the Road*
From the beginning, we felt organic farming must represent economic sustainability for farm families. The expression "Y in the Road" comes from our foundation bookkeeping system, which puts the total income into the farmer pool account. The farmer's target price is taken out first and then the balance goes to the business. A regular cooperative does the opposite; it pays the business expenses and then gives the farmers whatever is left over! *(pages 42–43)*

* *Character*
Defining and maintaining character is critical as the business ages and grows. CROPP Cooperative has always been full of character (and characters). A consultant once described our character as "quirky with integrity." In the midst of a debate in one of our early meetings, when everyone was is close, then we feel there has not been enough deliberation or we don't yet have enough information. At root we believe that reasonable people dedicated to a common vision can work together. *(page 88)*

* *Supporting the individual*
While consensus is our goal, we have always been aware that in order for a business to be dynamic, individuals need to be able to make decisions when they have to. Whenever this happens, it is good to sit down afterward with a larger group to acknowledge and review the circumstances that led to this action. By understanding and supporting the decision, we increase trust and encourage an entrepreneurial attitude among employees. *(page 180)*

COOPERATIVE / DEMOCRACY

* *Cooperative vs. corporation*
The purpose of a corporation is to maximize the value of the business even if that means selling the business. The purpose of our cooperative is to serve our family farm members fairly today and into the future. Our farmers' dream is that their children will take over their farms and that this cooperative will be there to serve them as we are serving the present generation. We don't have to face selling our business. Our mission is all about serving organic farmers for generations to come. *(pages 37, 39)*

getting exasperated, one of the farmers, Tom Forseth, stood up and said, "What would a regular co-op do? I say we do just the opposite!" (*page 39*)

* *Organic*
Organic is about learning from nature. It's the art of seeing every situation from a holistic view that looks for the cause and effect of the issue at hand. Organic is our foundation theme. It governs all aspects of our business. We have explored what it means to run a business organically and how we treat our employees organically, and so on. Organic is also about having faith that, using this philosophy, we can solve our problems. (*pages 32, 39, 55, 193–194*)

CONSENSUS

* *Group mind*
CROPP Cooperative has long believed in people working together. Meetings, meetings, meetings. Yes, a full cooperative process may take longer, but in the end we have built a foundation for the future and have made a wiser decision. (*pages 88–89*)

* *Chaotic consensus*
We have no obligation to consensus, but in our decision-making process, we have always strived for a broad buy-in. During our chaotic conversations, folks often feel that we are not getting anywhere. Then suddenly a consensus forms from the chaos. This experience gives us great faith in group process. (*page 34*)

* *None of us are as smart as all of us*
This idiom has become a foundational tenet for us. No matter how well someone thinks something through, it always seems that someone else can think of a new perspective. (*page 15*)

* *Voting and not enough information*
CROPP leaders and members vote regularly, but we strive never to have close votes. If a decision

↗

* *Democracy*
We often remember Winston Churchill who said, "Democracy is the worst form of government, except for all those other forms that have been tried." Surely a democracy has shortfalls and we have learned of some of those as we've matured. Elections can be popularity polls instead of leadership selection. Decisions can be made for political reasons, not long-term mission and business reasons. Individuals can overwhelm organizations by volunteering for every committee. We have learned from these lessons and have instituted policies and structure to overcome these shortcomings. Our nomination committee screens Board of Directors candidates to assure that only the best two candidates are on the ballot for each open seat. We have also appointed committees where possible; we find we get much better representation. Such controls can be abused but represent antidotes to the flaws of democracy. (*pages 181–182*)

* *Farmer-run*
A point of pride for our farmers is that the cooperative is "farmer-run." But it's really not the whole story. We are a farmer-owned and -governed cooperative, but the employees run the day-to-day affairs. The staff (many of whom are farmers) provides well-thought-out proposals so the farmers can make the wisest decisions possible. This relationship between the farmers and employees is a foundational partnership that has built our unique organization. Understanding governance will be an ongoing educational and policy issue for our cooperative. (*pages 89–90*)

LEARNING ORGANIZATION

* *They won't let you*
In the beginning our cooperative had a big vision. We knew that we might not succeed, but we were determined to try. We heard, "they won't let you" and "you all are really playing out of your league."

↙

We trusted our vision and backed that faith with hard work. We never did figure out who "they" were who were not going to let us do what we did. Another saying we have followed advises, "Those who think something can't be done should get out of the way of those doing it." *(pages 60, 127)*

* *Common sense and fairness to all*

We were started by farmers, so we were blessed with a wealth of farmer common sense. We also had the foundation of *organic,* which, at its root, is about balance; to us this means fairness to all. This fairness foundation is closely aligned with the deep root in the farm community of the Golden Rule. *(pages 10, 13, 80, 173)*

* *School of hard knocks, or the school of experience*

The leaders of our cooperative had little experience in running such a business but were willing to do their best, learn from their mistakes and experiences, and keep getting better. *(pages 26, 36)*

* *Compromises*

When and what to compromise is a challenging question for a mission-based cooperative. Some values we would never change, like being 100% organic or to never sell below our target prices. Others we have compromised as we have faced business and market realities, but none were made without deep discussions and awareness. For example, we have shipped huge amounts of freight to places where we didn't have a large regional farmer presence. This helped us avoid shortages and develop markets. Later, we followed up by starting a regional pool of farms to supply locally. For the sake of sustainability, we would rather not ship that initial long-range freight, but the compromise allows us a temporary, stable bridge to localize the production in the long run. *(page 110)*

↓

CONSTITUTION / DNA / MODEL

* *Formula thinking*

CROPP Cooperative's success is due to its consistent formula that has guided us through challenging times. Our formula is not a mathematical equation, but a mix of values, character and objectives that make up the foundational thinking of our business. A business faces a lot of changes along the way. It is critical to decide what is changeable and what is not. One sure test of a formula is whether it works at any size. We have learned that most of the time, alignment is alignment no matter what size. Our foundational principles are written to describe and define this formula. *(pages 62, 110, 119, 122, 142)*

* *A working model*

CROPP Cooperative has provided a model for the farmers to join that supports them in the brave move to organic farming. The very existence of a stable, supportive, economically-sound alternative has shaken agribusiness, which has always been used to being the only option. *(pages 33, 60, 62, 173)*

* *Writing a constitution*

It is critical to write down your DNA, or constitution, so that it is clear to all participants. At CROPP Cooperative we use our mission statement, goals, definition of organic and foundational principles to communicate our winning formula to future stakeholders. *(page 62)*

* *Fear of bureaucracy*

As we have grown, we have developed systems and processes. How to do this without becoming too bureaucratic is a big question. Hazardous, self-serving bureaucracy is hard to recognize, because it is always clothed in efficiencies, or order, or better processes, which are all necessary refinements

↓

* *Sound business*

From early on, we knew to listen to our consumer-partners, and their purchasing intent had to dictate our product line. Organic consumers wanted to buy an 8-ounce-sized cheese so we needed to have lots of packaging. They wanted calcium in their orange juice so we added calcium. We learned that to have a mission-based cooperative, we first had to run a sound business with good business practices and good customer service. A mission or vision is not valuable unless it can be transformed into a viable business. *(pages 84, 116, 146, 148)*

* *Role of a champion*

Despite our collective roots and team management style, we learned early on the importance of identifying a champion to lead a task. Someone needs to be the point person to make group process work. More importantly, one needs a champion with personal passion for the project. Finding the combination of facilitation and dedication in a leader is a critical factor in our success. *(pages 102, 139)*

* *Proposals lead the way*

The best leadership comes from someone making a proposal. A well-thought-out proposal quickens group process with supporting facts and logic. Guided by proposals, group process can be an efficient way of doing business. *(pages 123, 181)*

* *Dynamics of dualism*

Dynamic tension is the natural state of effectiveness. Tension is not a negative word when it acknowledges different priorities that complement each other in a productive way. As CROPP Cooperative grows, we face dualisms as a learning organization: we are bureaucratic and flexible; we go boldly forward in a cautious way; we care for farmers and employees; we are both a national and a regional cooperative. *(pages 17–18)*

when applied in balance. Avoiding self-serving bureaucracy has to be a focus. Otherwise, a business can easily lose its light-footedness, suffer from increased overhead, or break the entrepreneurial spirit of its employees. *(page 181)*

DIVERSITY

* *The 3-legged stool*

A foundational marketing strategy of our cooperative is to have diverse sales. The three principle components are retail packaged products, bulk sales, and ingredient sales. This diversity of sales helps the cooperative to better balance its members' perishable raw production with the market—a challenge that CROPP Cooperative faces on a daily basis. *(page 122)*

* *Dedication to a common vision*

Diversity of people has been a key component in our success. We have proved that dedication to a common mission will transform that diversity into a foundational strength. The combination of "traditional farmers" with "back-to-the-landers" with "professional folks retreating to the country" provides the broad perspectives that have been a major factor in our success. *(pages 33–34, 96, 127)*

* *Family farm*

As CROPP Cooperative expands around the country, we have discovered the diversity of American family farms, from the Mennonite or Amish 30-cow farm to the traditional Western 300-cow farm. This journey from region to region, exploring what a family farm means, is one diversity challenge we have embraced. *(pages 5, 10, 177)*

PEOPLE

* *Culture*

It is critical for an institution to decide what kind of culture it wants to nurture. At CROPP Cooperative we nourish tolerance and acceptance of diversity in both our farmer culture and our employee culture. This is the true pulse of the Cooperative. *(pages 80, 153, 172–173)*

* *Human factor*

CROPP Cooperative has always talked about the human family and how many unique stories we represent. We have realized that to be fair, we have to consider these many stories. "One size does not fit all" is a truism that we have experienced and practiced. We have tried to bring compassion into our cooperative culture, realizing that all humans have flaws and unique situations. Indeed, we joke that all would be fine if not for the human factor. In accepting our differences and our flaws, we are able to work together to serve our common mission. *(page 175)*

* *Gibb's Triangle*

One of our farmers introduced us to the Gibb's Triangle, a business-philosophy that clearly shows two paths. The first begins with trust in human goodness, then sharing common goals, then learning to communicate while minimizing rules and controls. The second path is just the opposite; rules and controls are the dominant mechanism, and trust is minimized. Over the years this contrast has been a vivid reminder for our cooperative. We have chosen the first path of broad trust and avoided the temptation to control everything human. *(page 89)*

↓

PARTNERSHIP

* *Co-dependency*

CROPP Cooperative started with a dream but no resources. Since organic was such a new thought, we had to find a way to work within the existing agriculture infrastructure. We were fortunate to have support from conventional agriculture; our initial supporters were our local tobacco cooperative and, most notably, the National Farmers Organization (NFO), who became our first milk handler. This beginning led us to a foundational tenet of building our business based on a co-dependency of resources. We hired someone to haul our milk, someone to make our cheese and someone to cut it into retail packages. We have been blessed with excellent partners ever since. Co-dependency has its risk, but the advantages clearly outweigh the vulnerability. This way, we avoid the risk of owning our own plants, yet we are free to pioneer new markets. *(pages 52–53)*

* *Conventional agriculture*

We were originally ridiculed for our idea to farm organically. But we spoke of organic as a different set of choices and avoided conflict with our neighbors so that we could remain part of our community. Our success was dependent on a good relationship with the conventional agriculture community, both in the present infrastructure and as our future source of new organic farmers. In the beginning there was tension, but that eased as organic became an economic force. Today the tension is again steadily increasing as science methodically reveals the dangers of chemical agriculture to human heath. Our cooperative has a duty to speak the scientific truth about the dangers of chemical agriculture and of genetically engineered foods. *(pages 32, 136)*

↓

* *Finger pointing*

The urge to blame one another is a fundamental human weakness. We love the old saying that every time you point a finger at somebody you have three pointing back at yourself. CROPP wants its employees to face their conflicts head on, to reach out to each other to resolve them. This is a responsibility of being an employee and should be part of every job description. To carry grudges, talk at the watercooler or whisper in the hallway without solving these conflicts is not earning one's wage. Yes, this is idealistic and not possible in all cases, but a philosophy we strive for. *(pages 180–181)*

* *Employees and trust*

Following the lessons of Gibb's Triangle, CROPP Cooperative provides a work environment that allows employees flexibility to develop a unique schedule and yet get the job done. If employees have children in soccer games, we want to allow them to attend as long as they make sure the job is covered. This practice and philosophy of two-way trust helps us retain talented, hard-working employees who fit the CROPP culture. *(page 89)*

* *Employee fit*

Many times we have seen excellent employees struggle because they are not in the right role. Moved to another role, they thrive and become more valuable employees. What a difference when we finally get a round peg in a round hole. Often we joke and ask an employee "what do you want to do when you grow up?" This is not entirely a joke. Everyone wants to feel valuable and contribute. An important role of every supervisor is to find the right role that allows each employee to blossom. *(page 181)*

↗

* *Competitors*

Organic has always been a pioneering effort that was more movement than business. Plenty of healthy competition exists within the organic industry, but we share the same inspiration and community. We all believe that there is plenty of opportunity. For CROPP Cooperative, a major organic dairy competitor did not show up until we were almost 5 years old, when Mark Retzloff and Paul Repetto, founders of Horizon, came to our door to purchase organic milk. We debated enabling a competitor, but not for long, since we had faith in the future of organic dairy and knew we needed help in developing the category. Retzloff and Repetto were experienced in the natural foods industry and had the skill set we did not have. The success of Horizon was a major force in the explosion of organic dairy into the number-one category of the organic marketplace. Mark Retzloff always said, "a rising tide floats all boats." *(pages 84–85, 147, 149)*

FUN FACTOR

* *Lighthearted*

Meetings at CROPP Cooperative have never been boring because jokes are the norm. Part of that lightheartedness is being able to speak truthfully but not offensively. When an organization is learning by mistakes, you'd better be able to joke about it. *(page 177)*

* *Social fun*

The employee culture always included social events, from holiday parties to canoe trips to after-work gatherings. The excitement of our work in fulfilling our mission carries over into becoming a family of comrades. To this day we still share fun when we can, whether it is our ping-pong and euchre games or the many employee social circles. CROPP farmers also have developed their own social events with the advent of pasture clubs, field days and regional meeting picnics. *(page 177)*

↙

* *Barefoot*

Jerome McGeorge and George Siemon, founding volunteers and employees, love being barefooted. Early on, there was plenty of joking, but out of it came the tradition of employees going barefoot in the workplace. Now it's common to see employees barefooted as they go about their job—it's become a metaphor for the acceptance of individual choices among CROPP employees. *(page 172)*

* *Employee wellness*

The workplace of CROPP Cooperative has blossomed with wellness programs for the employees. This fostered a culture that has changed peoples' lives with improved health, diets and exercise regimes. Focusing on an employee's well-being, whether physically, emotionally or spiritually, will make the employee a happier and more productive person. Wellness programs have produced real savings in our health insurance cost. Our employees even get a discount on insurance if they enroll in a CSA (Community Supported Agriculture). *(page 174)*

* *Farmer wellness*

An issue we have not addressed enough is wellness programs for our farmers. Farmers may farm organically, but that does not mean they have adopted an organic diet or lifestyle. The co-op subsidizes purchases of the products we produce to make them available for all members no matter where they live. We provide information and articles on organic cooking and more. We know there is more to do; this is a challenge we have yet to face fully. *(page 174)*

↓

* *Focus*

Business books always talk about the importance of focus—as they should. Focus, though, is not a simple conversation. Certainly CROPP Cooperative's focus on organic and not facilities has been a wise discipline. Diversifying into our many pools could be challenged as losing focus, but really the issue is not variety as much as synergy. Expanding into soy or orange juice fits into our focus of managing fluid processing plant relationships, similar quality programs, our refrigerator distribution system and our relationship with the refrigerated grocery buyer who is the same buyer for dairy. Our meat business, Organic Prairie, was not within that focus and struggled to succeed. Meat had different processing plants, had different quality concerns, required frozen distribution and presented a totally different buyer. Even so, meat was a good fit to provide a market for dairy farmers' excess animals. And after all, animals are a key part of a diverse organic farming system.
(pages 142–143, 193)

* ***Roughly right or exactly wrong, analysis paralysis and other planning ills***

CROPP Cooperative has always been an intuitive organization. As we have matured we have developed more planning systems but not without conflicting with our intuitive side. Somewhere along the way we heard the phrases "roughly right or exactly wrong" and "analysis paralysis," and we try to remind ourselves of the importance of balancing raw data with gut instinct. Some have said that employees with advanced business degrees often struggle working at CROPP Cooperative because we do not go at business in the same way as the textbooks advocate. Chasing the last decimal point in a plan could be the most important fact to determine or it could be irrelevant; telling the difference is the real challenge. *(page 150)*

↓

BUSINESS THEMES

* *Build the business then the building*

CROPP has always tried to minimize investment in facilities. One of the lessons we learned from our partnership with NFO was that cooperatives often lose their mission as they become beholden to the "brick and mortar" in their manufacturing plants. When CROPP bought its first processing plant at Chaseburg, Wisconsin, we quickly learned how true this could be. We saw how one decision might be good for the cooperative overall but cause losses in the processing plant. Our philosophy is to build the business using co-pack plants or rented warehouses or offices until the need is well established and then consider building our own facility. (pages 114, 125)

* *Plans for a new venture: poor, OK and super*

Paul Repetto of Horizon once told us, "If you are going into a business you should have a plan and a willingness to accept the results if the business did poorly, if it did OK, or if it was wildly successful. If you are comfortable with the risks and results of any of these scenarios, then you can afford to enter that business." The foundation of a good business plan is multiple assumptions. (page 143)

* *Business plans*

Business plans are necessary tools but assumptions can change regularly, so building a long-term plan on them can be hazardous. An old saying tells us that when you assume, you can make an *ass* of *u* and *me*. We have often joked that the wisest course in following a business plan is to divide the projections in half and multiply by zero. This is our unique reminder to ourselves to also rely on advice from others and our own intuitive sense, as well as the business plan. (page 143)

↗

* *Too big*

After our first 5 years of strong business growth, CROPP farmers and employees started asking if we were getting too big. This is a great question, because it leads to the question that matters most: "What is more important: size or mission?" Often a farmer would show concern about getting too big and would challenge us. In response, we would ask that farmer if they had a neighbor who wanted to join the Co-op. Inevitably the farmer would say yes, and detail how they had worked on their neighbor for years and how badly that farmer needed the financial stability CROPP could offer. Our reply was clear, "Then you want the co-op membership to double in size—that's what taking on your neighbor means when you realize all CROPP farmers are working on their neighbor like you are." This got us back to discussing the core question of being true to our mission. So as we get bigger, we need to ask ourselves, "Are we being true to our mission?" If we are not, then it's time to stop growing until we regain our mission focus. (pages 175–177)

* *A social experiment disguised as a business*

As a successful business, we often hear that folks think we are just another corporation profiting off of the organic movement. But being part of CROPP is so much more than a business. We have many missions: saving family farming through providing a sustainable market, providing employees with meaningful work, proving that partnering is better, and connecting consumers to good organic food from real farmers. In interacting with our stakeholders, listening to the stories of our farmers, hearing the passion from our consumers and seeing our decision-making always center on our values, we feel the movement we are all part of. So in truth we are a social experiment disguised as a business. (page 173)

7. A Social Experiment

Disguised as a Business

2007–2013

* Supporting the individual * Resisting bureaucrac

* Trust, initiative, fit * Democracy * Too big? *

Friends joked that Colleen Skundberg and Mary Shird were going to work with hippies when they joined CROPP in the early '90s. "We had an open house to celebrate our new retail store and the event may have turned some thinking around," says Shird. CROPPies welcomed the townsfolk from around La Farge, Wisconsin, and they served delicious organic food. They looked like regular people, after all.

CROPPies who joined the Cooperative in earlier days received red bandanas, says Skundberg. They fit the "dress code" which was—and still is—ultracasual: jeans, denim skirts, t-shirts, work boots, flip-flops or bare feet in summer. Some CROPPies come from local farms, having tackled the morning chores before reporting to work. Even the CEIEIO goes barefoot around the office, and it's common to see people barefoot as they go about their work. The only guy sometimes in wing tips and a tie is CFO Mike Bedessem, but that's what bankers expect.

Over 25 years, CROPP has developed a unique and sometimes quirky Co-op culture that works for them. Amidst this diversity of dress and backgrounds, CROPP people share a commitment to preserving and expanding organic farming in the face of a crumbling national food system that is more concerned with profit than people's health.

Years ago, when CROPPies were scattered up and down Main Street in La Farge because the old creamery was no longer big enough, it seemed that every rented office had a coffee pot and a toaster. When she joined CROPP to manage creative services, Sarah Bratnober was assigned a buddy. "Helen Jo Zitzmann showed me the ropes and introduced me to people," Bratnober says. "In every little office, they were making toast. Or cheese sandwiches." It was simply the neighborly thing to do.

David Bruce, manager of CROPP's juice, eggs, meat, produce and soy pools (JEMPS), remembers climbing creaky stairs in a dark hallway at the old creamery. "I was there for a job interview and the first person I saw was Jerome McGeorge, shirtless with his hair in pigtails," says Bruce. "He cracked a big smile and shouted, 'Welcome!' That was my introduction to CROPP. I thought, 'Oh, yeah, this could be home.'"

> "Folks are walking around in shorts and they don't have ties, but they're dead serious; they know where CROPP is going and they're determined to get there."
> Jerry McGeorge

Don't Be Fooled

That casual, homespun atmosphere has sometimes been misunderstood, says Jerry McGeorge, CROPP's director of Co-op affairs. On his first day, McGeorge remembers being "impressed by the camaraderie and strong sense that we were all together in this amazing experiment." McGeorge says, "I later realized that even though CROPP has a casual veneer, people have total dedication to the mission and they work really hard. I've watched some people be fooled by the veneer. Before we moved into our new headquarters, some visitors looked around and let surface impressions lead them to believe CROPP was some fly-by-night organization."

Like its farmer founders, today's CROPPies are attracted to the Cooperative's mission. "Why am I here?" says McGeorge. "I want to make the world a better place, and I think CROPP offers a great opportunity to do that."

> "I love the idea that I'm working for farmers in my community—down-to-earth people—not some suit behind a desk."
> Kasey Larson, Chaseburg office and warehouse supervisor

← Some of the many faces of CROPP Cooperative

A Social Experiment

In recent years, when CROPP was in the news for its brand popularity and rapid growth (about 20 percent annually), George Siemon described the Cooperative in a way he never had before. "We're a social experiment disguised as a business." It struck a chord. The description accurately reflected the Cooperative's business model and its organizational culture.

> "We try to walk our talk and live the values we believe in. Today, that's still incredibly unique in the business community."
> Theresa Marquez, CROPP chief mission executive

When it was founded, CROPP rejected agricultural norms that emphasized mega-production, chemical fixes that depleted the soil, and ag cooperative practices that failed to put farmer-owners first. Instead, CROPP opted for a different business model that emphasized a fair and sustainable pay price for family farmers and respect for the earth and its creatures.

When times were especially good (and profitable), suitors knocked at CROPP's door. Its farmer-members chose to remain independent and prepare their enterprise for the next generation. CROPP refers often to its "triple bottom line": people, planet and profits. "We're not doing business the way other businesses are doing it," says Theresa Marquez, CROPP's chief mission executive. "A two percent profit is enough for us. A fair wage for farmers is enough for us. We're not looking to sell out and become rich."

When he talks to finance people, Mike Bedessem describes the CROPP culture as unselfish. "We had people urging us to sell," says Bedessem. "Some people said, 'Do it because it's in the best interests of the farmers. Maximize their value.'" But those were minority voices. "We weren't seduced by the money that was flowing around us," he says. "Instead, our farmer-members wanted to keep the Cooperative independent." Nor did CROPP do what some co-ops have done: limit the number of farmer-owners and keep the financial rewards for a few. "That idea was rejected when CROPP had 50 farmers and it was rejected again when we had 100 farmers," Bedessem says. "I call that unselfishness."

In poor and drought-prone countries, organic systems already produce 80% more than conventional systems.

Asked to describe CROPP's unique culture, people offer many adjectives, but these come up most often: quirky, casual, resilient, deliberate, collaborative, humble, run like a family business, compassionate,

← CROPP staffers (from left): Steve Gundlach, David Bruce, Dan Hazlett, Mickey Keeley (front, in hiding), Josh Lewison and Jim Kojola in 1994; (from left) Dawn Parr, Laura Potter, Tammie Lee and Mickey Keeley in 2010

Mike Bedessem in 2011 →

curious, fun, hard-working and mission-driven. With rapid growth and the aging of CROPP's "founding fathers and mothers," the Cooperative is putting heavy emphasis on preserving this culture and preparing the next generation to carry it forward.

Humanity at Work

Back in 1988, Jerome McGeorge, CROPP's first CFO, encouraged CROPP's leaders to learn about organizational culture from the Mondragon Corporation, a federation of worker-owned cooperatives founded nearly 60 years ago in the Basque region of Spain. Today, Mondragon is the country's seventh largest company and its theme is "humanity at work." Their federation of more than 250 companies emphasizes participative leadership, solidarity, business excellence and respect for people and their lives outside of work.

From CROPP's organic cafeteria (the Milky Way Café & Kitchen), outdoor vegetable garden and yoga classes to its flexible hours, wellness incentives and support of volunteerism, the Cooperative has adopted "humanity at work" as its own cultural norm. An employee growth incentive (EGI) program, created by Theresa Marquez and Jerry McGeorge, rewards people with cash if they better themselves. That might mean a new exercise program, losing weight, learning a new skill or volunteering in the community. Should CROPP be paying people to do things they ought to be doing anyway? "It's a source of debate," says Jerry, "but we want to create an environment where we say, 'We value you as people and we want you to be the best you can be. This is one of the ways we're going to help you do that.'" EGI bonus payments arrive in paychecks around mid-December, when employees always welcome a little more cash for the holidays.

A survey of 800,000 workers at 300 companies found stress-related sick days tripled from 1996 to 2000.

CROPP's wellness programs have produced real savings in the Cooperative's health insurance costs. Employees also get a discount on their insurance if they take part in a Community Supported Agriculture (CSA) vegetable program, or join a local gym. Now it's time to emphasize wellness programs among farmer-members, CROPP leaders say. Burning calories in the field each day isn't enough to ensure good, all-around health.

We Got Family

People speak of CROPP as a family and that's literally and figuratively true. Several next-generation sons and daughters of CROPP farmer-members have joined CROPP's employee base. At one point in its evolution,

← Kickboxing class at CROPP headquarters

Highway clean-up volunteers in 1994 (from left): Sally Marshall, John Capehart, Rebecca Zahm, Jerry McGeorge and Zell Spry; next page: CROPP's gardening class in 2009 →

about 10 percent of CROPP's employees were married to each other. "The best benefit by far was meeting my husband, Grant, when he started working at Organic Valley in 2001," says Helen Jo Zitzmann. "We were married on New Year's Eve 2005."

Co-workers watched Harriet Behar and Aaron Brin fall in love as they collaborated on an organic butter pamphlet for CROPP. "That pamphlet," says Colleen Skundberg, "turned into a love letter."

> "I've grown to think of my co-workers as my extended family." Paco Rondeau, Cashton Distribution Center accounting

When Nichole Benson joined CROPP more than a decade ago with a degree in public relations, she thought the job would be "just a stepping stone." Today, she's a regional dairy supervisor working with a network of processing plants and dairy haulers. "I have two little girls at home," she says. "CROPP really cares about all of us and our families." Long-time observers say CROPP has brought compassion into its cooperative culture, recognizing that a single work schedule or policy does not fit all people. Employees and farm families have unique situations and this "human factor" is given considerable weight.

> "I support the organic mission in my life as a consumer and farmer. I am constantly reminded of how interconnected we are." Karen Hermsen, event properties coordinator, CROPP II Warehouse

Mission-Minded

Most CROPP employees point to the mission as the prime attraction in joining the Cooperative, though not all arrived in La Farge as avid organic consumers. They are like many of CROPP's farmer-members who chose organic farming for the financial security of a sustained pay price, but as they learned more about organic, they became advocates. The excitement of work to fulfill a shared mission has carried over into creating a "family of comrades" at CROPP.

David Bruce saw the early stages. "Everybody poured their lives and their passions into the enterprise," he says. "Then and now, we're something very different and I feel like I can be a part of carrying our mission forward."

Too Big?

With more than 1,800 farmer-owners in the U.S. and Canada and more than 600 employees running an enterprise with sales approaching $1 billion in 2013, its 25th anniversary year, this "social experiment" has become

← This butter pamphlet brought Harriet Behar and Aaron Brin together.

Sisters Heather Christianson (left) and Heidi Davidson (right) both work at CROPP and their parents, Sandy and Kenny, are CROPP farmer-members. →

a big business. Nearly 100 processing plants help CROPP produce 1,470 individual organic products for the consumer and ingredient markets. Altogether, CROPP moves 3.5 million pounds of organic goods daily. It is the largest organic dairy cooperative in the U.S., providing 50 percent of the organic milk nationally and about 60 percent of the organic milk in California, alone.

> "We all share the same mission and that is always putting farmers first." Jeff Kragt, senior director of production and manufacturing

"But is CROPP getting too big?" some farmers and employees have wondered. That question prompts a more fundamental question: "What is more important, size or mission?" If a CROPP farmer wants his friend or neighbor to join the Cooperative, then it could grow by one more farmer, right? But what if hundreds of CROPP farmers encourage their neighbors to join, as many do? How big is "too big" then? As it stands now, CROPP will continue to grow as organic farmers need them and consumers want their products, as long as the organization remains true to its mission.

> "A lot of people would like us to control our growth. But when farmers come to us and want to join CROPP, it's hard to stick to a strict growth plan." Eric Newman, vice president, sales

As it has grown, the Cooperative sees the amount of good it can do, but if CROPP strays from its mission, it's time to stop, take stock, and regain the mission focus.

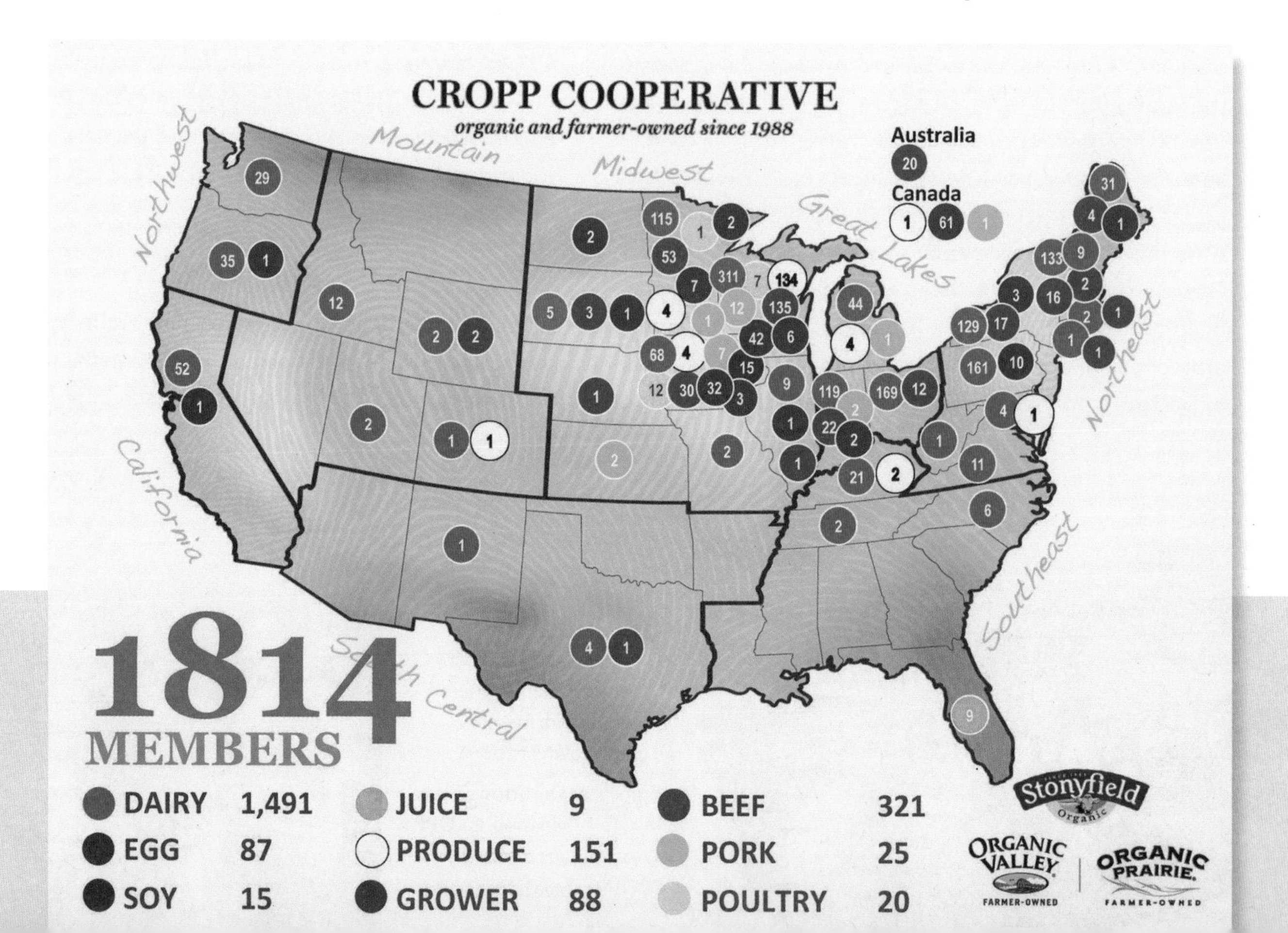

Asking the question 'How big is too big?' keeps the leadership on their toes and accountable.

The same "too big" question has been leveled at the size of individual farms. As CROPP expanded its membership to many geographic regions, the definition of a family farm was not the same everywhere. A Mennonite/Amish dairy farm in Ohio or Indiana might have 30 cows, while a traditional Western farm herd averages 300. Some organic family farms in California even have dairy herds as large as 1,500. CROPP has had to expand its own definition of a family farm to recognize this geographic diversity. At its core, CROPP defines a family farm as one owned and operated by families with a focus on protecting the land and supporting the community and the rural economy for future generations.

Staying Connected

The days of CROPP's dairy pool members meeting in the back of Phil and Deb's Bar in downtown La Farge or the staff gathering around a few picnic tables are long gone (though picnic tables still grace CROPP's expansive HQ "front yard.") Now CROPP's extended family connects through conference calls and lots of travel. All-staff "P&Ws" (once known as Pow-Wows) and CEO Chats occur monthly to keep employees in the loop at headquarters, and they are broadcast via computer to employees in other locations. Employees also have an opportunity to give the Cooperative feedback through the Cultural Council, an idea espoused by Jerome McGeorge and inspired by the Spanish Mondragon business model. "It was designed to be a voice for employees and a channel for feedback," says Jerry McGeorge. "Employees elect their representatives, and their charge is to deal with things that relate to our culture. It's fairly unique, so it fits into the category of social experiment." In addition to cultural topics, large and small, the Cultural Council is a kind of focus group. "It gives us a chance to feel the pulse of our employees," says Jerry. "We tell them about an idea we're considering and ask them what they think."

After hours at CROPP, some still gather for "Thursday night Therapy," a tradition started years ago. There are holiday parties, canoe trips on the Kickapoo River, ping pong and Euchre games. CROPP's farmers have developed their own social outings with pasture clubs, field days and regional picnics. The well-attended CROPP annual meeting is a big family reunion (babies included) with lively reminiscing, celebration of the combined success of employees and farmers working together, reports on the state of the Co-op and abundant organic food.

The Highest Paid Dishwasher

As it has grown, the Cooperative has reached outside its locale for talent and beefed up its financial offers (the days of everyone earning $5 an hour seem like fiction

← Here's to good barbeque: CROPP Cheese Room team members chow down in 2006.

At the 2009 annual meeting, CROPP attendees took a turn at basket-weaving. →

IT Never Sleeps

Embracing the high-tech world of Information Technology (IT) was not a pretty sight in 1996 when Randall Juenemann signed on with CROPP to be the Cooperative's first IT guy *and* communications coordinator.

"Our data center was in the observation room above the cheese-cutting room in our first headquarters," Juenemann says. "Every once in a while the drain in the floor would back up with this God-awful oozing liquid. We put down a 2 by 4 plate so we didn't see it."

CROPP had about 30 computers running Windows 3.1 in 1996 and had just adopted Novell 3.11 networking, an advanced system at the time. Far less advanced was CROPP's accounting system stored on Peachtree, an off-the-shelf application for small businesses.

Things started to happen fast as CROPP's revenues (and its attendant data storage and handling needs) spiked from $13.8 million in 1996 to $72.6 million in 2000. John Capehart, an early computing whiz, joined Juenemann in 1998 to begin designing the architecture for CROPP's IT system, and in 2000, George Neill left a much bigger company in La Crosse to lead the department (today called Information Resources).

"Our 'data center' was a bread rack," says Neill, "and when the Co-op outgrew our headquarters and we rented office space up and down Main Street in La Farge, we borrowed the city's line truck to string fiber cable ourselves." By about 2002, the IT staff, totaling some ten people, moved to a white trailer outside HQ nicknamed "Beluga."

Proper cooling for the IT hardware was always an issue in those trailer days. "Our consultants visited and they were sweating bullets," says Curt Parr, who joined the CROPP team in 2003. "We said, 'We're building this new building and it'll have a proper data center with racks and spacing and cooling, but for now, we have to get by with this.'"

Construction of CROPP's new "headquarters on the hill" in 2004 gave IT their biggest challenge: choosing a partner and implementing its first Enterprise Resource Planning system designed to serve all CROPP departments well into the future. "We're not a traditional food manufacturer," says Neill, "so the majority of our folks are knowledge workers who sit at computers. They communicate with customers and vendors and distributors and consumers all over the country."

CROPP's IT pioneers remember how fast the requests for money grew. "I made a case for a $500 expenditure around 1997 and that was a large amount then," says Juenemann. The next big request was $68,000 and then $254,000. By 2013, CROPP's IT budget hovers in the $15 million to $20 million range for a system that handles four times more data than all the books in the Library of Congress. CROPP's IT language has turned to talk of "virtualization capabilities," cloud computing, public clouds versus hybrid clouds and something called the "virtualization hypervisor"—language that only an IT pro can truly understand.

CROPP was an early adopter of the World Wide Web, too. In his dual role as communications coordinator, Randall Juenemann designed CROPP's first website around 1998 to help support CROPP's mission, build its brands and educate consumers about the benefits of organic foods.

When mIEKAL aND arrived in 2000 to serve as CROPP's first "webmaster," digital marketing concepts hadn't even entered the picture for Web users. "CROPP was focused on farmer support and strategy, not technology," he says. But thanks to Theresa Marquez, CROPP's then chief marketer, Carrie Branovan and a cadre of tech-savvy folks, the Cooperative put Organic Valley out there in "cyberspace" well before the language of *hits* and *search engines* became commonplace.

By 2005, CROPP had three web "properties," as they're called: OrganicValley.coop where CROPP promoted its branded products, CROPP's mission, education and community involvement; OrganicPrairie.coop, devoted to promoting the Cooperative's branded meat products; and EarthDinner.org, a site designed to build awareness around the impact of food choices with family-friendly recipes, meal plans and conversation starters. CROPP's primary site, OrganicValley.coop, attracted about 500,000 people in 2005.

CROPP's web feature with the greatest "seniority" is "Down Nature's Trail," a journal by long-time employee and naturalist, Dan Hazlett. His thoughtful ruminations on living close to the land were some of the first e-blast messages that CROPP employees (who were a little leery of the whole website thing), looked forward to receiving on their computers.

Around 2006, CROPP added another website called Farmers.coop expressly for current and potential farmer-members who want details such as regional pay prices, recommended pasture sizes, product quality measures, information on organic livestock care and even a "clothing mootique," where farmers can "let the world know you're an Organic Valley farmer."

CROPP's Web pioneers have experimented with "micro sites" linked to informational campaigns such as CowsUnite.org, comparing the treatment of organic and conventional livestock; Sustain360.org, an online web forum dedicated to giving organic consumers a place to share ideas; OrganicRising.com, an organic breakfast site; and OrganicHeroes.com, an organic partnership site.

Today, when someone types the word "organic milk" into Google or another search engine, one of the first results is a link to Organic Valley's website. As many as 40 percent of the visitors to OrganicValley.coop find the site this way. Many are avid Facebook friends and Twitter followers, too.

Web and social media applications are part of virtually every marketing or educational effort that CROPP employs these days, says Nathan Lenz, a web developer from 2005 to 2012. "It is broadly available," says Lenz, "and here's the best part: it never sleeps."

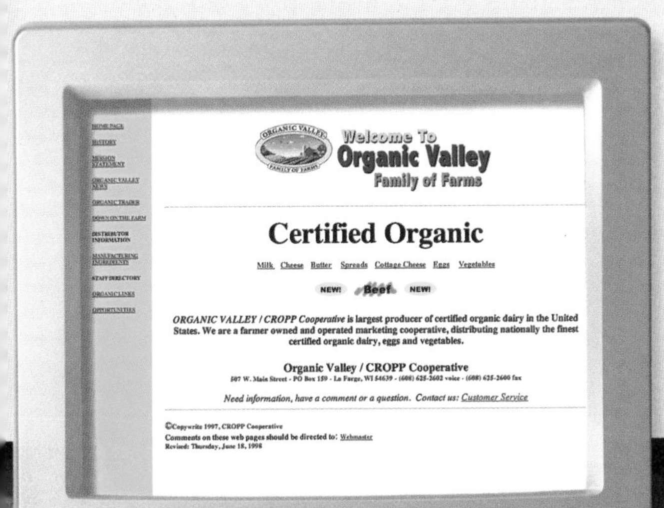

(Left to right): Randall Juenemann looks for a solution to the paper glut in 1996; CROPP was an early-adopter of Web communications; tech wizard Curt Parr and his son, Thane, in 2006

← CROPP staffers (from left): Rita Challet, Lisa Olson, Michael Tulley and Eric Snowdeal →

today). "Pay is still a challenge," says CFO Bedessem. "In a cooperative, pay scales tend to be more modest and the rewards come from supporting the mission. George Siemon likes to say he's the lowest paid CEO in an $800 million business and the highest paid dishwasher. We're just now paying market wages." (Employees also enjoy generous profit-sharing through contributions to their 401K accounts as a percentage of their annual pay, a program started in 2004.)

Relocating to a rural community can be a barrier to hiring, too, Bedessem says. It's not always easy for potential CROPPies to imagine themselves living in rural southwestern Wisconsin. Kelly Gibson, CROPP's key marketer for New York Fresh, its locally-produced, bottled and distributed milk, remembers her first interview in La Farge. "I left home in downtown Chicago and arrived in La Farge at midnight," she says. "It was a ghost town. I stayed in a studio apartment above George Wilbur's law office on Main Street. There was no key and no lock. I had the city mentality and I was terrified. I put a chair up against the door."

Gibson was already a loyal Organic Valley consumer, and the next day, she immediately liked the people at CROPP. But how could she leave the excitement of Chicago and settle in this all-too-quiet, rural burg? It seemed impossible. Theresa Marquez recruited Gibson heavily and the Chicagoan has never regretted her decision. "The size of my job doubled about a week after I got to CROPP," Gibson says. "It's been a fantastic personal and professional experience." When she wants a shot of nightlife or first-class performing arts, she spends a weekend in the Windy City.

> "There's a certain amount of chaos in our business and we believe it's OK. We can't control everything. It can drive some MBAs crazy." Eric Newman

Trust, Initiative and Fit

Over time, CROPP's leaders have taken their cooperative spirit ("Everything's better together") and translated that into a workplace culture that supports initiative, leadership from all corners, fairness, personal responsibility and trust. People are given a lot of freedom at CROPP and they are supported. They don't have to worry about losing a job over an honest mistake. "If our business is going to be dynamic," says George Siemon, "people need to be empowered to make decisions alone and under pressure, if necessary. Whenever this happens, it's good to sit down afterward with a larger group to acknowledge and review the circumstances. By understanding and supporting the decision, we increase trust and encourage entrepreneurial thinking among our employees."

When something goes wrong, the urge to blame one

another is a fundamental human weakness, but CROPP is clear about wanting employees to face their conflicts head on, reach out to each other and resolve issues without finger pointing. Carrying grudges or whispering in the coffee room without solving a conflict is not part of earning a fair wage at CROPP.

Sometimes excellent people struggle at the Cooperative because they've been assigned roles that don't tap into their strongest interests or talents. In that case, CROPP expects supervisors to find the suitable "fit" for a person so he or she can blossom, even if that means a new boss and a new assignment.

> "As an organization, I think we're just beginning to emerge from our adolescent exuberance. We have begun to know and understand who we really are."
> Jerry McGeorge

The best leaders at CROPP take initiative, ferret out the facts and make concrete proposals that they believe are best for the organization as a whole. A well-thought-out proposal streamlines CROPP's sometimes lengthy, consensus-building group process. These proposals can come from any corner of the Cooperative.

Folks in charge are tuned-in to creeping bureaucracy, too. "We've developed systems and processes," says Siemon, "but the big question is how do we incorporate these practices without becoming too bureaucratic? Sometimes it's hard to recognize, because bureaucracy can be clothed in more efficiencies, order or better processes, which are all necessary refinements. Avoiding self-serving bureaucracy has to be our focus, otherwise we can easily lose our light-footedness, suffer from increased overhead or harm the entrepreneurial spirit of our farmers and employees."

Distilling Democracy

As CROPP has matured, it has faced some difficult times at the Board of Directors level because the Cooperative has always believed in the democratic process. George Siemon likes to quote Winston Churchill. "No one pretends that democracy is perfect or all-wise," Churchill told the House of Commons in 1947. "Indeed, it has been said that democracy is the worst form of government, except all the other forms that have been tried . . . "

From its inception, CROPP has adhered to the International Cooperative Alliance's 7 Cooperative Principles, and this one is clear about the democratic model: "Cooperatives are democratic organizations controlled by their members who actively participate in setting policies and making decisions. The elected representatives are accountable to the membership. In primary cooperatives [which CROPP is], members have equal voting rights, one member-one vote." This

← Farmer-members at CROPP's 2011 Annual Meeting (from left): Arne and Carol Kleppe-Gonvick, Glen Harder, Roger Peters, and Alden and Chad Christianson →

is true no matter how small the farm. It empowers these smaller family farmers and it negates an old cooperative practice that allowed high-producing farmers to dominate cooperative decision-making.

Though all boards of directors are expected to govern and look out for the best interests of the organization they serve, human leadership can fall short. CROPP learned, through hard experience, that board elections could become popularity polls instead of leadership selection. Decisions might be made for political reasons, and some directors could be self-serving. Directors could also overwhelm the board process by volunteering for too many committees. To counteract these problems, the CROPP Board formed a nominating committee to screen board candidates and recommend the best two candidates for each open board seat. The Board also appoints committees when they can, rather than simply accepting volunteers. Even these practices can be abused, but CROPP leaders believe these common-sense policies help minimize the potential flaws of democracy.

In response to a burgeoning local foods movement, the New Oxford American Dictionary chose "locavore" as its word of the year in 2007.

Moments of Truth

There was no minimizing the flaws in the marketplace during the late 2000s, however. The severity of the supply problems and economic pain tested CROPP's mission and resolve. By 2008, the organic milk gold rush was over. People who believed there could never be too much organic milk on the market saw the opposite: bloated inventories, over-supply problems, fierce competition, sinking retail and pay prices. In addition, because there had been a big influx of new organic producers, organic feed was in short supply.

The year 2009 and the years that followed proved that CROPP could maneuver through dramatic swings in market conditions and come out whole. It reminded CROPP that there were many outside factors that required agility and frequent tweaking of their plans.

CROPP was careful to keep its bank debt to a minimum so that its financial footing was strong. To the surprise of many, the Class E, Series 1 Preferred Stock that CROPP began offering in 2004 had become an unqualified hit and a stabilizing financial force by the late 2000s. Class E stockholders own an equity interest in CROPP without the voting rights, and they earn a cumulative annual dividend of 6 percent, paid quarterly. Class E investors include people of all stripes, regions and ages. Most have never farmed, but they believe in CROPP's mission and its desire to remain independent. Many make a point of visiting CROPP during its annual hosting of the Kickapoo Country Fair at the end of July for the annual shareholder's meeting. Over six years, from 2004 to 2010, these shareholders invested more than $40 million in CROPP. Remarkably, there were $7.5 million in new investments in 2009 when America's stock market took its precipitous dive. In 2010, more than $14 million in additional funding came in to CROPP by September, rendering the program so successful that CROPP then closed its sale of stock. Despite the volatility of the economy that left many other companies struggling to survive, CROPP simply didn't need more money.

Finding Feed

Though funding was solid, feed was scarce. CROPP's farmers needed feed to supplement pasture grazing, especially in northern regions. If they could buy organic feed at all, the cost was usually prohibitive. Some farmers returned to raising their own feed and reducing their animal numbers to achieve a sustainable balance on their farms.

In the face of this shortage, CROPP launched its own organic feed program so that farmer-members had a ready, affordable source. At first, the program simply connected a farmer in one region to buy or sell to a farmer in another region. In addition, CROPP used its significant buying power to purchase large quantities of feed for re-sale to its members. That program spawned CROPP's newest pool—a grower pool—in 2008.

In its first year, CROPP signed up 23 growers from 15 states with more than 3,000 acres of corn, soybeans, milo, wheat, barley, oats and hay. CROPP asks its farmer-members with cattle, hogs or chickens to lock-in three-year feed orders for the amount they would normally purchase off their farms. By doing this, the farmers can count on stable, sustained feed prices and the feed growers can expect the same. It's all about mutual support. Fluctuating organic feed supplies and prices continued to be a headache, however—so much so that CROPP would increase its pay price by $2 per hundredweight (100 pounds of milk) in 2012 to help farmers manage spiked feed costs.

After building a $17.5 million primary warehouse and distribution center in Cashton (in 2007), installing a new software system to track supply and demand (2007), increasing product prices even in the face of over-supply (2007), seeing its Midwest dairy pay price reach an all-time high ($24.75 in 2008) and starting a brand new grower pool (2008), in 2009 Cropp faced its first annual loss since the early years. Though the loss was small—1 percent—the basically break-even year was a shock to CROPP's 1,404 farmer-members and 512 employees who had grown accustomed to hearty growth year after year. Fortunately, CROPP had a strong financial foundation, high quality farmers in place and long-term management to cope with an economic era they had never before seen: the longest and deepest global recession in eight decades. They vowed to get through it and come out stronger in a cooperative, organic way.

Cooperation and Solidarity

The Cooperative froze hiring in 2009 and CROPPies proved they could still keep their positive attitudes and endure while running a lean business.

The oversupply problem was harder. As one observer said, "We had a standing policy on how to increase production, but none fully developed to decrease production!" It wasn't the first time CROPP faced oversupply issues, but never on this scale. "We were already oversupplied and yet our base program allowed our dairy farmers to produce 34 percent *more* milk," says Jim Wedeberg, CROPP's dairy pool director. "Our base

← Gregory (left) and Greg Brawner of Hanover, Indiana, members of CROPP's Grower Pool

We Walk Our Talk

Maged Latif was a single guy when he came to the United States from his home country, Egypt. With a degree in geology and chemistry from the University of Cairo, he had been a high school teacher in Jordan. He came to the States to visit a friend and explore life here.

Latif married an American woman whose family had immigrated to the U.S. from Jordan, and he found work with a food company while he earned his M.S. in food science at the University of Illinois. Latif became CROPP's first Director of Quality, Research & Development and Certification in 2002.

"In my work before joining CROPP, I formulated premium food products to compete with national brands," he says. "Then in the 1990s, I watched many in the American food industry shift from premium formulations to lower cost alternatives. There aren't as many ingredient choices when your goal is lower cost, but we were successful in doing it. We cheapened the food products. Bottom line, the foods we formulated weren't something I'd give my kids to drink or eat."

At the same time, says Latif, health scientists were discovering the dangers of some of these lower cost ingredients, such as high fructose corn syrup and artificial sweeteners. They have since been linked to obesity, cardiovascular disease, diabetes and cancer.

Latif found OrganicValley.coop while surfing the Web, and he was attracted to CROPP's mission and the promise of cleaner, healthier food. "I started reading about organic and I asked people at my company about it," he says. "Even the scientists had little understanding. They mixed it up with organic chemistry."

"... there is nothing more important than food—how we raise it, how we distribute it and how we consume it. At a time of rampant obesity, especially among children, nutrition should be a national policy." Columnist Kathleen Parker, August 3, 2012, *The Washington Post*

When he was hired at CROPP, Latif asked COO Louise Hemstead about the Cooperative's research and development lab. "She took me to the kitchen and said, 'this is it.' We actually did some bench research in that kitchen, but we also rented pilot plants to do some of our initial testing."

Latif's office was a small corner of a trailer outside CROPP's retail

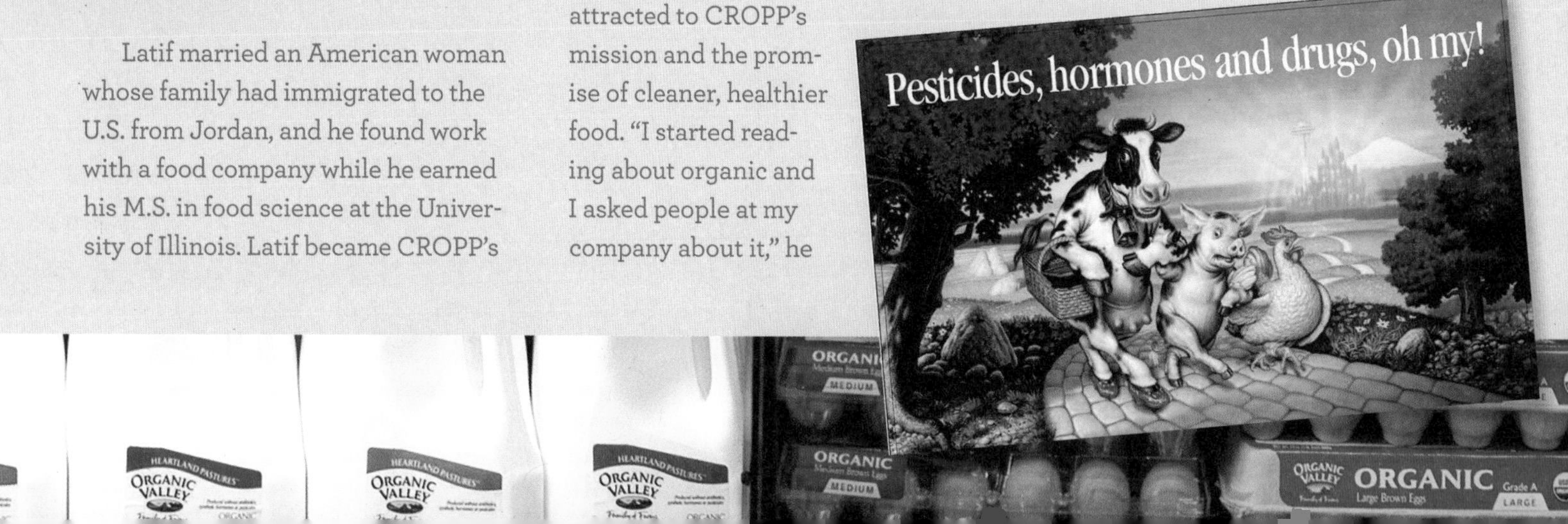

store in La Farge, quite a come-down from his former 25-by-25 work space. "I had been treated very well and paid very well," he says, "but I didn't feel right about earning that money because I didn't feel good about the work I did."

Today, Latif supervises four different departments at CROPP: quality assurance, product development, packaging development and certification. He also has a new R&D pilot lab in CROPP's newly-expanded headquarters in La Farge.

Having seen the American food industry from both sides, Latif is convinced that organic practices are vital if we are to repair the damage done by agricultural chemicals used heavily in America. "Look at cancer," he says. "It's showing up at even younger ages. That is a consequence of what we've been eating." In a 2006 study of children, ages 3 to 11, for example, scientists found that children eating conventional diets had pesticide levels about six times higher than the organic group. (For more scientific studies, visit organic-center.org). Researchers have also discovered that certain chemicals and hormones found in conventional livestock feed cause cows to produce milk with high levels of omega-6 fatty acid, a substance that increases the retention of unhealthy fat in the human body. Organic milk does not have this excess omega-6. This is an asset, given the rise in obesity and its proven link to cancer. This comes from Dr. David Servan-Schreiber, author of *Anti Cancer: A New Way of Life*, published in 2009.

Latif continues,"Organic practices are all about not harming our soil with chemicals, the proper uses of water and thoughtful care of animals. I know that our products are created for a sustainable future. Consumers need to know that our products are superior and made with care and love reflecting our true mission. I have seen CROPP in action at close range for a decade. We walk our talk."

COCKTAILS, ANYONE?

Most Americans are exposed to a cocktail of 4 to 7 endocrine disrupting agricultural pesticides on a near-daily basis. Numerous studies have shown that a diet of organic foods significantly reduces the amount of pesticides in children who previously ate conventionally produced foods. Organic Valley farmers never use synthetic pesticides, herbicides, or fertilizers, ensuring food that's as safe as it is healthy and delicious.[7, 8, 9, 10]

COWS ON GRASS

Many studies conclude that cows getting 30% or more of their daily food from pasture produce milk that is higher in protein and conjugated linoleic acid (CLA), a heart-healthy fat that also plays important roles in a child's development. The milk from pastured cows is also higher in the Omega 3 fatty acids that are lacking in our grain-heavy diets today. That's why Organic Valley farmers uphold the most rigorous pasturing practices.[11, 12]

WE DON'T TAKE 'CIDES

Organic Valley and Organic Prairie farms don't use any synthetic fertilizers, pesticides or herbicides. A recent study based on USDA data compares average use of these toxic substances on conventional farms with our cooperative organic dairy farms over the past 20 years. Every gallon of Organic Valley milk avoids the use of approx. 1 gram of synthetic pesticides and herbicides and 2.5 oz of synthetic nitrogen fertilizer.[13]

program was way out of kilter, so we went back to the drawing board." The solution was to revamp the base program, monitor it month by month and temporarily limit supply in order to keep pay prices strong.

> "The new system was a watershed moment. It proved we could operate effectively even when demand drops quickly." Wayne Peters, board president, 1992–1993, 1998–1999 and 2002–2010

For the first time in the Cooperative's history, the organic dairy pool implemented a mandatory supply quota for its farmers in July, 2009. Using a three-year production average as an "active base," each farmer was asked to limit his or her supply to 93 percent of that base. Any production over the percentage earned the farmer $15 per hundredweight (100 pounds) less than the full organic pay price. The dairy pool quota ended in August, 2010 and that year's sales were so strong that CROPP's farmers received more than they gave up when the year's accounts were settled. They even earned an unexpected "13th check" bonus. Even so, dairy pool members hoped the quota would gather dust on a shelf before market conditions called for using it again.

CROPP's other pools looked for ways to reduce production in 2009, too. Some pools—notably the egg pool—made decisions to lower their pay prices to cover the cost of selling their organic eggs on the lower, conventional market until consumer demand came back (fortunately, a few farm operating costs were temporarily lower: feed and fuel). "The egg pool executive committee came to the CROPP Board and said, we want to reduce our pay price by ten cents a dozen because the pool is losing money and we want to be responsible to the whole Co-op," Jerome McGeorge remembers. "That's what I call cooperation and solidarity." As a group, CROPP's farmer-members proved to be solid leaders willing to sacrifice income to safeguard CROPP's long-term goals.

> "CROPP has always tried not to act like we know all the answers." Cecil Wright, director of operations and employee support

The Right Thing to Do

Humility is part of the CROPP ethos and it has been since that rag-tag bunch of idealistic farmers got together in 1987 to create the Cooperative in the depths of America's farm crisis.

But it may have been a little tough maintaining that humility when the phone call came from Hood, the dairy company founded in 1846 with this advertising tagline: "Always Good. Always Hood." Hood had

Margie and Lavern Martin Farm in Decker, Michigan; David and Robert Petersen on their Humboldt County, California, farm →

joined the organic milk market in 2004 in partnership with Stonyfield Farm. Hood licensed the use of their brand for organic fluid milk and competed fiercely with CROPP.

However, Hood, in 2009, notified CROPP that it was exiting the business. Hood had a proposition: Was CROPP interested in acquiring the organic milk production of 280 organic farmers?

Jerome McGeorge remembers the offer. "We were on a three and a half day strategic planning retreat in La Crosse and all we talked about was Hood," he says. "At the end of the retreat, we had 'group mind' agreement and we said 'yes,' even though it was a bigger gulp of milk than CROPP had ever swallowed at one time. It was the right thing to do for these family farms."

From 1988 to 2008, CROPP farmers kept 90.75 million pounds of synthetic pesticides, herbicides and nitrogen fertilizer out of America's soil and water.

During the fall and early winter of 2009, CROPP's leaders visited every farm in the Stonyfield Group to describe the Cooperative's mission and its approach to organic agriculture. They checked the organic integrity of the farms, especially their pasture practices. They told farmers they could elect to join CROPP or simply finish out their contracts with Hood. In the end, all but 20 of the farmers joined CROPP on January 1, 2010. They started first in a "reserve status" while the Cooperative dealt with the considerable logistics of adding so many new farmer-members at one time. Within nine months, the farmers were full members of the CROPP dairy pool, earning the same premium, stable pay price that everyone else did.

In addition, CROPP had a new processing partner in Hood, more efficiencies in milk hauling and an even stronger relationship with Stonyfield (with their beloved brand, Stonyfield fluid milk). There was another benefit, too, says Eric Newman, vice president of sales. "Not only did we have two dairy brands to deploy where the brands were strongest, but we earned more credibility in the eyes of grocery retailers. The industry thought, 'Organic Valley is taking over. Look at that. They're players now.'"

Other organic dairy groups also joined CROPP in 2010 and 2011, including most members of Wisconsin Choice, Lancaster Organic Farmers Cooperative in Pennsylvania, two organic dairy groups in Oregon, and North Coast and Humboldt County Dairy Farmers of America, both in California. When they joined CROPP, these farmers immediately saw a $2 to $4 pay increase per hundredweight for their organic production.

The decision to take on all these new farmers may confound some observers of CROPP, says Newman. "Some people think that we don't have a strategy that we follow," he says. "But every year, we put goals and initiatives out there. In 2011, we had to make careful budget adjustments when we took on all that extra milk. We had to do it to protect the farmers, but it was a huge opportunity, too."

"Light-Footed" and Energy-Conscious

When it came to investments, CROPP's commitment to construction of a $17 million, 80,000 square foot distribution center in rural Cashton, Wisconsin, was the Cooperative's single largest expenditure in its history. (There are already plans underway for an expansion at Cashton, as well as construction of an office building on the site.)

↑ Interior of the Cashton Distribution Center and the Cashton Logistics team in 2008; the grand opening at Cashton included a talk by then Wisconsin Governor James Doyle (second from right at ribbon cutting); a second 2.5 megawatt wind turbine was installed at Cashton in 2012; Cecil Wright, CROPP's director of local operations and sustainability, with Jeff Rich of Gundersen Lutheran Health System of La Crosse, Wisconsin

"When we make decisions, we're very deliberative. We ask: 'Why are we doing this? What's the risk? Can we contract it out?' We lease and then we build."
Louise Hemstead, chief operating officer

In its deliberative style, CROPP's leaders took time to make this big investment in bricks and mortar, or more accurately steel and concrete. Existing space was available to rent, but CROPP decided to design and build its own distribution center. About three years of thinking and planning preceded groundbreaking. CROPP looked at locations from Madison to Minneapolis, but settled once again on a tiny rural farming community of about 1,000 people some 15 miles up the road from La Farge.

"We wanted to have a facility that could handle all of our specialty products so a customer had the convenience and efficiency of a one-stop shop," says Louise Hemstead, COO. CROPP's fluid milk was already being produced, processed, bottled and delivered in its key regions (as were most Organic Valley eggs). However, CROPP's many specialty products, often with a longer shelf life, could best come from a single, central distribution center, rather than eight leased warehouses. "We designed Cashton to be light-footed. Every item we offer is there and ready to go out," she says.

In her travels, Hemstead examined distribution centers and warehouse floor plans, quickly discovering that vertical storage (with more floors and automated retrieval using cranes and forklifts) would minimize the building's footprint and make it more efficient and safer. That has proved to be the case: the whizbang Automated Storage and Retrieval System (ASRS) that moves mega-quantities of products every day at Cashton even injects energy back into the power grid.

← A 33,000-foot addition to CROPP's headquarters in La Farge was completed in 2012.

The Koester dairy farm in Illinois uses solar panels to generate power. →

The Cashton Distribution Center is positioned on a 40-acre spread that is part of Cashton Greens Industrial Park, an innovative new development where businesses will utilize renewable energy from wind turbines. In 2011, in partnership with Gundersen Lutheran Health System, of La Crosse, Wisconsin, CROPP began installation of two, 2.5 megawatt wind turbines there. Up and running in May, 2012, the wind project is expected to produce about 6 million killowatt-hours per year—enough clean energy to offset more than 80 percent of the electricity used the previous year at Cashton, CROPP's Chaseburg Creamery and its La Farge headquarters combined.

The man behind these green and sustainable initiatives is Cecil Wright, a former business systems analyst for Nationwide Insurance in Columbus, Ohio, whose real love was making and selling organic maple syrup. "I was a foodie," says Wright, "and I'd go to some of the same shows where Organic Valley participated." A friend introduced him to George Siemon and Mike Bedessem and he joined CROPP in 1998. Wright worked first in IT and then with Louise Hemstead in operations.

Sustainability had always been part of the CROPP vocabulary, and Wright was attracted to ways that the Cooperative could be an even better environmental steward. Trucking products all over the country consumes a lot of fuel, after all. There had to be other things CROPP could do to help offset that and contribute to environmental sustainability. For example, when CROPP expanded its La Farge headquarters in 2011, the 33,000-square-foot addition was based on green design principles and met LEED certification standards (LEED stands for Leadership in Energy and Environmental Design). The addition incorporates solar collectors on the roof and in its window systems, and uses solar energy to heat water. There are free-standing solar collectors beside the headquarters parking lot, too. CROPP's hauling and employee vanpool fleets use increasingly higher averages of biodiesel blended fuel. More CROPP farmers are conducting energy audits on their own farms and some are even growing oilseed for fuel.

Offshore Fans

Helping to fuel CROPP's growth—sales reached $715.6 million in 2011—are an increasing number of consumers who live outside the U.S. For example, while about four percent of America's food dollar is spent on organic, that percentage has swelled to nearly 10 percent in Europe, says Harriet Behar, CROPP's original product sales dynamo (she later became an organic certification inspector and director of MOSES, the Midwest Organic and Sustainable Education Service).

In fact, Behar is credited with making the initial contacts to sell Organic Valley cheese in Japan in the 1990s. CROPP later began selling its products to Houston-based American Grocers, supplier to most U.S. embassies and American expats around the world. Next

came sales to DeCA, the company that stocks U.S. commissaries at military installations worldwide.

"We made a concerted effort in the late 2000s to invest in export and international markets with our branded products," says Eric Newman, CROPP's vice president, sales. One La Farge staffer, Curtis Olson, is dedicated to international sales, and Pete Bassett travels internationally selling organic meat and dairy products. CROPP has also engaged a Chinese citizen full-time to build sales in that country. "There's a deficit of organic pork and dairy products in Asia," says Newman. "Mexico, the Caribbean and Central and South America are growing organic markets for us. We don't sell much in Europe because that region produces so much already."

> "We're sending refrigerated milk from New Jersey to Argentina in five days. That's about the same energy consumption as transporting across the U.S. by truck." Eric Newman, vice president, sales

Some observers have questioned CROPP's interest in international distribution of its products, but Newman offers one example of why it makes sense. "Let's say we're packaging our organic milk in aseptic packages that have a nine-month shelf life and don't require refrigeration. Put that milk on an ocean-going container ship and it takes less energy to reach China than it does to move milk cross country to the eastern seaboard."

Even a soldier in Iraq can buy Organic Valley milk, and "we often get letters thanking us," says Newman. In fact, on-line purchases of CROPP's products in the U.S. and abroad will soon become commonplace. Organic Prairie frozen products are now being sold on-line through Costco.com and Organic Valley's less perishable products will soon join them. Before long, consumers will be able to purchase many Organic Valley products on-line from its own website.

What began 25 years ago as a bold experiment in a little corner of rural Wisconsin reaches, quite literally, around the world today. The Midwestern farm crisis gave CROPP its start, but the Cooperative continues to grow dramatically because of a worldwide consumer revolution focused on the quality and safety of food.

> "The inspiration of the organic movement is contributing to a better world. It's easy to have faith in that." Jerome McGeorge

The Promise of A Better Way

It was a steamy July evening in La Farge, Wisconsin. They gathered at The Kettle, a rustic hilltop home and gathering place for CROPP's first generation of leaders, perched high overlooking the Kickapoo Reserve and its meandering river, rolling hills and lush hardwoods.

They talked about the future, seated at the same age-nicked dining table where Jerome McGeorge drew up an astrological chart for the fledgling venture born of America's farm crisis.

Weighty problems of start-up and survival were long past, replaced by new questions of size, succession and legacy.

Twenty-five years ago, CROPP had three visions for its future.

Jerome McGeorge: *First, we had a commitment to the organic movement; second, we were dedicated to breaking out of the conventional dairy world and its gyrating prices to establish stable and sustainable prices for family farms; and third, we wanted to be the best ag cooperative.*

← Charlene Stoller with her daughter, RoseMary, on their Ohio dairy farm

"There's a Native American truth that says past and future are always with us." Jerome McGeorge

CROPP retains these same three visions today, but the stakes have gotten higher. Nearly 2,500 organic farm families and employee families depend on CROPP for their livelihoods. A growing number of investors have put their faith in CROPP. A complicated web of independent milk haulers, freight carriers, processing plants, organic food manufacturers, distribution centers and warehouses bring CROPP's products from member farms to millions of consumers in the United States and beyond.

CROPP is one big "social experiment disguised as a business," but how much bigger can it—or should it—get? The same can be asked about family farms, which vary dramatically in size.

Theresa Marquez: *There's a big debate about what is a family farm. There are 50-cow farms and there are farms with 1,000 cows or more.*

George Siemon: *What unites them is their relationship to the land, to the places they farm and to their communities. If we can pay them a sustainable pay price, they can stay at home . . . and really enjoy life.*

Some organic farmers are still lonely out there. They might not have organic-minded neighbors, but CROPP gives them kinship, through membership, with more than 1,800 organic farmers like themselves.

"The beauty of a family farm is having a relationship with a place. Family farmers are happy being there, in fact, they're the luckiest people on earth." George Siemon

As it grows, CROPP's geographic regions will become stronger entities, in part because the "live local, buy local" philosophy has gained traction. Thirty years ago, before CROPP was founded, the bioregionalism movement in the Midwest echoed today's thinking of local interdepedence, but it was ahead of its time.

Theresa: *CROPP's future is in developing our regions. They will grow and have their own identities. Each region will be very different in terms of farm size, how they apply organic practices and their community involvement. And yet, all the regions will still need each other—perhaps more. Think about the extreme weather we're facing. Our climate problems will move around the United States and we're going to have to help each other.*

George: *Mentoring young organic farmers in our regions will be important, too. We'll look into other services we can offer our farmers that can assure them a secure, holistic lifestyle. Along with organic farming education, maybe it will include financial planning, insurance or estate planning.*

About four percent of the American food dollar is spent on organic today, and it's not unrealistic to imagine that percentage growing to ten percent in the next decade. Are there enough organic farmers in America to produce that food?

↑ Son Lorin Yoder, bringing in the cows on Kore and Miriam Yoder's family farm in Lewisburg, Pennsylvania

(From left) Sam, Worth and Louisa Johnston on their family's farm in Tillamook, Oregon ↗

George: *No. CROPP has grown organically since its origin, but it will take more organization and money to bring new farmers into organic farming. We're doing things we never imagined, like helping investors buy land and rent to organic farmers. Our Cooperative is investing in more bricks and mortar. We will have to deeply invest and ask CROPP farmers to be willing to sacrifice for the community, because that's our mission.*

CROPP is also focusing on succession and preparing the Cooperative's next generation of leaders.

Theresa: *By 2025, we're going to have a whole new set of leaders and George, Jerome, myself and others will all have retired. Selecting and preparing them will be our biggest challenge.*

George: *We have to keep on doing what we're doing and do it well, so that CROPP is there for future generations. The good news about us is that our farmers already know who they want us to be in 50 years. They want us to be an honest vehicle that serves their ventures.*

At the beginning of this book, we posed a basic question: Can America (or the world, for that matter) be fed without poisoning the land or the farmers who raise the food?

George: *Organic can feed the world without poisons. In fact, we should feel morally obligated to do so. Right now, we have too many acres devoted to producing corn, fructose and sugar. At the same time, we have big ag claiming that only technology can feed the world. How can depleting our natural resources through destructive technology be a solution?*

Theresa: *My question is: Why are organic advocates always asked that question? Why are we the only ones developing sustainable, long-term ag solutions, while big ag pursues a model focused on profit first? My question is: How do we fix a broken food system? Why are people starving and undernourished and what can we do about it?*

Ask any thoughtful observer of CROPP's evolution and that person will come away convinced that this Cooperative has contributed significantly to fixing America's broken food system. CROPP has created an economic engine that makes organic production realistic and sustainable, but it was pioneering farmers and consumers working together that made it successful.

Theresa: *The organic food industry has always been a consumer-driven and a farmer-driven movement. That's what I've loved about it.*

↑ Janet Baker on her family farm in Enumclaw, Washington

"All of us need to grow wiser in our ecological consciousness. I believe food is a big part of that." Jerome McGeorge

George: *Organic is giving everyone the opportunity to have a new conversation around food and the relationship of the farmer to the market. I want to express thankfulness from the farmers to the consumers because, by discovering organic, you've changed farmers' lives. Farmers love farming, but the economic treadmill got them down. Organic has given them a new life. You can see it in the vibrancy and you can see it in the farm youth.*

CROPP's success is clearly linked to consumers' growing awareness of food and its impact on their lives and their health.

Jerome: *The greatest change in our human consciousness is that each of us is responsible for our own health. Ancient Tibetan wisdom identifies three sources of health: what we eat, what we drink and how we behave. Each of us is responsible for all three.*

Theresa: *The International Federation of Organic Movements follows what's happening in countries around the world. There are 100 different nation-states where people feel the same way about organic as we do. They believe there's a better way to grow food. It starts with eliminating chemical poisons and it continues with soil and water conservation. We're all connected and that gives me hope. When we chose to go organic some forty years ago, it wasn't strictly about eating organic. It was also about recycling, using less energy and alternative educational models. It was about honesty, integrity, wellness and kindness. Organic was not just a food choice, it was a lifestyle choice.*

George: *When I think back to organic in the 1970s, it was an explosion of interest that had many "threads" as Theresa describes them. Those threads have come together now into the fabric of a bigger movement that is going to explode. If I had to sum it up, I would describe the movement as people working together to make the world a better place.*

↑ (From left): William, Scott and Brian, holding sons Brock and Caden, on the Forrest family farm in Torrington, Wyoming

↑ Johanna Deal on her farm in Mount Vernon, Texas

Leaping into the future with Peyton and Liana Darling, on the Roxbury, New York, farm of their grandparents, Larry and Jeanne →

Our mission is to create and operate a marketing cooperative that promotes regional farm diversity and economic stability by the means of organic agricultural methods and the sale of certified organic products.

Acknowledgments

The seed was planted for this book when George Siemon, Jim Wedeberg and Jerry McGeorge formed the 25th anniversary book committee in the summer of 2010. They drafted Sara Bratnober and Andy Radtke to serve as coordinators of the project. Sarah and Andy helped the committee craft a proposal and interview several author teams. Finally in February, 2011, the book committee chose the team that has lovingly produced the book in your hands.

Pine and Partners, of Minneapolis, was chosen to write, design and produce the book. CROPP's Core History Book Team included Andy Radtke, Rebecca Zahm, Jack Pfitsch, Spark Burmaster, and Robert Clovis Siemon.

Carol Pine interviewed scores of CROPP farmers and employees and performed extensive background research. Between November 2011 and June 2012 she wrote the first draft of the narrative of this book. Following extensive feedback from the CROPP book team, Carol produced a second draft over the summer of 2012 and then a final draft was completed in August, 2012.

Cathy Spengler, Pine and Partners' designer, developed the layout concepts for the book. With the help of the team, she selected illustrations for the narrative from thousands of pictures in the CROPP archives. Her work also included three rounds of proof and refinement.

The index was created by Julie Folie. The book was printed by Friesens, with help from sales representative Elizabeth Cleveland and customer service representative Erin Enns.

Andy Radtke, CROPP's content manager, served as the editor and chief of the project. He worked closely with Carol to perfect the final draft of the book. He worked with Cathy and Clovis organizing CROPP's photo resources for use in her layout. He also helped Jack with the editing of the essays and the "Threads of Wisdom." His steady hands on the reins focused the efforts of the team toward a quality product.

Jack Pfitsch, philosopher and one of CROPP's founders, worked closely with Carol on her research. He interviewed early CROPP employees and partners, adding to Carol's pile of transcripts. He coordinated the essay requests to prospective authors and corresponded with the volunteer essayists. He coordinated the editing of George Siemon's "Threads of Wisdom". He worked with Rebecca digitizing documents for the archives and creating a CROPP timeline. His "big picture" skills suited his role as "chief worrier" for the book.

Rebecca Zahm, CROPP archivist, coordinated access to current paper archives and the digitizing of CROPP's library of photos, artifacts and documents. She always managed to find the missing pieces needed when chasing down key facts. Her calm resourcefulness inspired our confidence that the answers to all our questions were at her fingertips.

Spark Burmaster, historian and one of CROPP's founders, culled through his own unique collection of historical photos from the beginning days of CROPP and worked with Rebecca digitizing CROPP's photos and archives. His attention to detail has been invaluable at every stage of the project. Many of the historical photos in this book are due to Spark's ever-present historical eye and his ever-present camera during the formative and anonymous moments of CROPP's dawning.

Clovis Siemon, filmmaker and CROPP's founding bystander (George's son), brought his knowledge of CROPP's stories and visual resources, and his media skills to this task. He worked tirelessly scanning photo archives and was a key resource for Cathy's layout. His youthful energy sparked the creative efforts of the group.

The original committee, George, Jim and Jerry, guided this project with productive meetings and helpful commentary. We have all been inspired to produce a book that honors their original vision.

Many people too numerous to mention helped our group again and again with information and leads to information. Farmers, employees and people touched by CROPP have been uniformly generous with their time and memories. Thank you one and all.

The people interviewed for the book: mIEKAL aND, Tony Azevedo, Jingles Badtke-Karuga, Jon Bansen, Pete Basset, Mike Bedessem, Harriet Behar, Scott Bennedict, Nichole Benson, John Boere, Bill Bosshard, Sarah Bratnober, Mike

Illustration Credits

Breckel, Greg Brickl, Aaron Brin, David Bruce, Spark Burmaster, Dan Casler, Paul Dettloff, Richard deWilde, David Engel, Marta Engel, Travis Forgues, Kelly Gibson, Diane Gloede, Jamie Gudgeon, Marcia Halligan, Al Hass, Ray Hass, Lisa Hass, Dan Hazlett, Louise Hemstead, Gary Hirshberg, Nancy Hirshberg, Aaron Hoover, Suzy Hultgren, Keith Johnson, Mark Kastel, Mickey Keeley, David Kline, Casey Knapp, Jim Kojola, John Kolar, Jeff Kragt, David Lathrop, Maged Latif, Willi Lehner, Nathan Lenz, Kelly Mahaffy, Lila Marmel, Theresa Marquez, Sally Marshall, Mark Martin, Ernest Martin, Jerry McGeorge, Jerome McGeorge, Dan Meyer, Mark Michel, James Miller, Jerry Miller, Matt Muellenberg, Doug Muller, John Nettum, Eric Newman, Steve O'Reilly, David Osterloh, Wayne Peters, Roger Peters, Jack Pfitsch, Jim Pierce, Reggie Pityer, Kitty Pityer, Mike Podesta, Carl Pulvermacher, Tom Quinn, Dave Randle, Dave Roberts, Clarence Rollins, Bert Roper, Charlie Roper, Brian Rude, Steve Ryssek, Pam Saunders, Mary Shird, George Siemon, Jane Siemon, Robert (Clovis) Siemon, Margaret Siemon, Doug Sinko, Colleen Skundberg, Lloyd Stueve, Nancy Stueve, Angela Swenson, Dean Swenson, Jan Swenson, George Teague, Adolf Trussoni, Arnie Trussoni, Beth Unger, Dave Vanderzyden, Greg Varney, Gloria Varney, Wayne Wangsness, Jim Wedeberg, Jake Wedeberg, Greg Welsh, Bill Welsh, George Wilbur, Guy Wolf, Cecil Wright, Rebecca Zahm, Helen Jo Zitsmann

Our thanks to the many CROPP farmers and friends who answered our call and provided various historical images for use in this book. Most we have attributed below, per page, to a photographer, but some, we regret, are from indeterminate sources and for which we are also grateful.

Cover:

Shawn Linehan: pasture photograph

Mixed Media: farm illustration for original Organic Valley logo

Book Pages:

Harriet Behar: pages 50–51, 55, 61, 115

Cheryl Walsh Bellville: inside front cover spread

Carrie Branovan: inside front and back cover spreads; pages 44, 69, 76, 98, 111, 113, 115, 116, 117, 118, 119, 139, 141, 142, 148, 149, 154, 170–171, 186

Larsh Bristol: pages 12, 44, 82, 99

Spark Burmaster: inside front and back cover spreads; pages 14, 24–25, 28, 30, 31, 33, 35, 36, 37, 38, 40, 47, 53, 55, 57, 64, 65, 66, 67, 68, 82, 84, 85, 89, 90, 92, 93, 94, 98, 171

c12/Shutterstock.com: pages 178–179

Anna Campbell: page 112

Rick Dalbey/Livengood Advertising: inside front and back cover spreads; pages 142, 145, 149

Dempag Entertainment: page 173

Elenamiv/Shutterstock.com: pages 20–21

The Engel family: page 16

Helen Jo Gudgeon: page 137

Mary Gundlack: inside front cover spread

Leah Harb: page 193

Lisa Hass: page 11

Joni Kabana: inside front cover spread; page 111

Jingles Karuga: inside front cover spread

Jim Klousia: pages 170-171, 174, 175, 179, 180

Shawn Linehan: page 193

Ernest Martin: page 120

Jerome McGeorge: page 55; all hand-written display elements throughout

Nastco/iStockphoto: pages 74–77

National Farmers Union archives: page 53

David Nevala: inside front and back cover spreads; pages 8–9, 80–81, 94, 101, 108–109, 119, 126, 134–135, 142, 146, 147, 170–171, 173, 190, 192, 193, 194, 195

Pastel Dakota Photo/iStockphoto: page 45

patpitchaya/Shutterstock.com: notebook for "voices"

Leon Reed: page 27

Mary Shird: pages 66, 128, 173, 194

Drew Shonka: page 175

Robert Clovis Siemon: inside front and back cover spreads; pages 64, 96, 152, 153, 179, 183, 184, 185, 186, 188

Colette Skundberg-Radtke: pages 93, 177

Otmar Smit/Shutterstock.com: page 126

Eric Snowdeal, III: inside front cover spread; page 188

Olaf Speier/Shutterstock.com: pages 138, 139

Ron VanZee/Successful Farming: page 125

Weston Imaging Group: page 154

Eric Wuennenberg: pages 170–171, 181

Ray Yoder: page 84

continued on next page

"voices" Portrait Photographs:

Carrie Branovan: pages 19, 129

Dregne Family: page 19

David Kline: pages 20–21

Connie Falk: page 45

Steffi Behrmann: page 47

Carol Pine: page 71

Kay Lewison: page 103

David Nevala: page 105

Kevin Newhall: page 131

Shane Opatz/Eau Claire Leader-Telegram: page 155

Karen Benbrook: page 157

Factoid Sources:

Page 13: DeMarco, Susan "A Fresh Crop of Ideas." *The Progressive*. Jan 1989. Vol. 53, No. 1, p27, 30.

Page 18: Sullivan, Walter. "First cloning of mammals produces three mice." *New York Times*. 4 Jan 1981. http://www.nytimes.com/1981/01/04/us/first-cloning-of-mammals-produces-3-mice.html

Page 32: Smith, Gregory J. and U.S. Department of the Interior, Fish and Wildlife Service. "Pesticide Use and Toxicology in Relation to Wildlife: Organophosphorus and Carbamate , Compounds." All U.S. Government Documents (Utah Regional Depository). *Paper 510*. 1987. http://digitalcommons.usu.edu/govdocs/510

Page 39: Lawless, Greg. "Humble Beginnings." University of Wisconsin Center for Cooperatives, *Bulletin No. 2*. Aug 2002. http://www.uwcc.wisc.edu/pdf/bulletins/bulletin_08_02.pdf

Page 43: "1988 Economy / Prices." *1980s Flashback*. http://www.1980sflashback.com/1988/Economy.asp. Accessed: 14 Nov 2012.

Page 43: "History of the federal order class III price (simple average)." USDA Upper Midwest Milk Marketing Area, *Federal Order No. 30*. Minneapolis Market Administrator's Office. http://www.fmma30.com/ClassPrice/HistoryofClassIII.pdf

Page 59: "Industry statistics and projected growth." Organic Trade Association. June 2011. http://www.ota.com/organic/mt/business.html

Page 62: "U.S. sees more risk of cancer in farmers." *New York Times*. 29 Sept. 1992. http://www.nytimes.com/1992/09/29/news/us-sees-more-risk-of-cancer-in-farmers.html

Page 65: "Appendix table 1—U.S. milk production, 1950-2000." USDA National Agricultural Statistics Service. Various years. www.ers.usda.gov/media/488851/append1_1_.xls

Page 83: Benbrook, Charles, et. al. "The first 13 years of GE crops consumer summary." *Impacts of Genetically Engineered Crops on Pesticide Use in the United States: The First Thirteen Years*. The Organic Center. November 2009. http://www.organic-center.org/reportfiles/13Years2-Pager.pdf

Page 88: Taylor, Stacy. "Fun facts about milk." 20-20Site.org. Love To Know Corporation. 2012. http://www.2020site.org/fun-facts/Fun-Facts-About-Milk.html

Page 97: Taylor, Stacy. "Fun facts about milk." 20-20Site.org. Love To Know Corporation. 2012. http://www.2020site.org/fun-facts/Fun-Facts-About-Milk.html

Page 114: http://www.leopold.iastate.edu/sites/default/files/pubs-and-papers/2003-07-checking-food-odometer-comparing-food-miles-local-versus-conventional-produce-sales-iowa-institution.pdf

Page 122: Shane, Scott A. "Startup failure rates—the REAL numbers." *Small Business Trends*. 28 Apr 2008. http://smallbiztrends.com/2008/04/startup-failure-rates.html

Page 125: "Background of Dairy Production in the U.S." U.S. Environmental Protection Agency. Last updated 27 June 2012. http://www.epa.gov/agriculture/ag101/printdairy.html

Page 140: Agriculture Fact Book, 2001–2002. USDA Office of Communications. p15. Mar 2003. http://www.usda.gov/factbook/2002factbook.pdf

Page 147: "History of the federal order class III price (simple average)." USDA Upper Midwest Milk Marketing Area, *Federal Order No. 30*. Minneapolis Market Administrator's Office. http://www.fmma30.com/ClassPrice/HistoryofClassIII.pdf

Page 150: Barrionuevo, Alexei. "Beef demand seen in peril as prices rise." *New York Times*. 10 June 2005. http://travel.nytimes.com/2005/06/10/business/10beef.html?_r=0

Page 173: "Organic farming could feed the world." Organic Consumers Association. July 2007. http://www.organicconsumers.org/articles/article_6657.cfm

page 174: "Workplace Stress." American Institute of Stress. Accessed 14 Nov 2012. http://www.stress.org/workplace-stress/

Page 182: "What is local?" *Sustainable Table*. Jan 2009. http://www.sustainabletable.org/issues/eatlocal/

Page 187: "Organic Counts! Organic Valley Launches First Online Calculator to Measure Personal Impact of Food Choices." Organic Valley press release. 23 Sept 2009. http://www.organicvalley.coop/newsroom/press-releases/details/article/organic-counts-organic-valley-launches-first-online-calculator-to-measure-personal-impact-of-food-c/

Index

Italics indicates an illustration